Nelson Williiam Perry

Electric Railway Motors

Their Construction, Operation and Maintenance

Nelson Williiam Perry

Electric Railway Motors

Their Construction, Operation and Maintenance

ISBN/EAN: 9783744678643

Printed in Europe, USA, Canada, Australia, Japan

Cover: Foto ©berggeist007 / pixelio.de

More available books at **www.hansebooks.com**

ELECTRIC RAILWAY MOTORS

THEIR

Construction, Operation and Maintenance

AN ELEMENTARY PRACTICAL HANDBOOK FOR THOSE EN-
GAGED IN THE MANAGEMENT AND OPERATION
OF ELECTRIC RAILWAY APPARATUS

WITH

RULES AND INSTRUCTIONS FOR MOTORMEN

BY

NELSON W. PERRY, E. M.

NEW YORK
THE W. J. JOHNSTON COMPANY
253 BROADWAY
1896

COPYRIGHT, 1894,
BY
STREET RAILWAY GAZETTE COMPANY.

CONTENTS.

CHAPTER		PAGE
	Preface,	v
I.	Introduction,	1
II.	Technical Terms, etc.,	9
III.	Ohm's Law; Rate of Work; Examples,	16
IV.	The Electric Current and its Properties,	28
V.	The Electric Current and its Properties (*continued*); the Solenoid,	39
VI.	Measuring the Current; Magnetism and Electro-Magnetism,	48
VII.	Lines of Force; the Closed Magnetic Circuit; Magnetic Leakage,	59
VIII.	Polarity, Magnetism and Current,	67
IX.	Electro-Magnetic Induction; The Continuous Current Dynamo; Increase of Electromotive Force; Eddy Currents in Armature,	75
X.	Shifting of the Armature Wires; Open and Closed Coil Armatures,	96
XI.	Drum and Ring Armatures; Consequent Poles and Multipolar Field Magnets,	102
XII.	Multiple Arc and Series Arrangement; Current Characteristics of Multiple and Series Arrangements in Generators,	109
XIII.	Current Characteristics in Translating Devices; Multipolar Fields,	116
XIV.	The Dynamo Electric Principle; Series, Shunt and Compound Winding; the Reversibility of the Dynamo.	129

CHAPTER		PAGE
XV.	The Electric Motor, ...	139
XVI.	The Electric Motor (*continued*); Torque,	145
XVII.	The Line of Commutation; Counter-Electromotive Force, ...	151
XVIII.	Counter-Electromotive Force and Speed Regulation; Requirements of Speed Regulation, ...	157
XIX.	The Sperry and Johnson-Lundell Systems,	169
XX.	The Leonard, Perry and Other Systems; the Perry System of Series Electric Traction; Storage Battery Traction; Conduit Systems, ...	177
XXI.	The Management of Street Railway Motors; Sparking at the Commutator; Motor Stops o. Fails to Start, ...	191
XXII.	Specific Directions to Motormen, . .	201
XXIII.	Instructions to Inspectors and Superintendents, ...	213
XXIV.	Locating Faults; Commutators; Drop Method of Testing for Faults; Insulation Test; Bearings; Gears and Pinions; Controllers, ...	220
XXV.	Trolley Wheels; Incandescent Lamps; Conclusion, ...	229

PREFACE.

The matter contained in this book first appeared as a serial in the *Street Railway Gazette*, beginning with the first issue in January and extending through succeeding issues until that of July 21. The ostensible object of the series was to provide for motormen an intelligent elementary exposition of the principles upon which are founded the apparatus which they, too often without any conception of their character, are called upon to handle. It has been assumed by the author that the audience is entirely ignorant both of the nomenclature and principles of applied electricity, but is anxious to gain a foothold, as it were, on the science. To this end he has avoided the use of all technical terms, except such as necessarily form the very alphabet of the science, and these have been met boldly—the principle being followed, however, of first explaining by simple experiments, such as are readily within the scope of a child of fifteen years, the phenomena themselves, and then assigning to these phenomena their proper names.

In the study of a new language, a new science or a new branch of mathematics it is the first principles that are almost always the most difficult to acquire. Few of us, probably, recall the task it was for us to learn the alphabet of our language. More of us will remember the difficulty experienced in acquiring the multiplication table, and all who have pursued the higher mathematics will remember how, after passing through algebra, geometry, and trigonometry, they were perplexed by the very rudiments of the differential calculus. As each step was taken the mind had to be prepared for the new study by means of analogies which, though never exactly true, enabled the mind to grasp ideas approximately correct only, but which served as stepping-stones to other ideas which approached more nearly the truth. In this work free use has been made of analogy, but by extending the analogies over a somewhat extended range, and by making use of some which are believed to be new, it is thought that the ideas inculcated by them will be as free from error as is possible under such circumstances.

The author has insisted upon a full understanding of Ohm's law—the fundamental law of the flow of currents—as a prerequisite to a further understanding of the subject. This law, however, is not sprung at once upon the student, but is led up to by the series of simple and inexpensive experiments before referred to, in the performance of which the student at last finds himself surprised that he has not only practically worked out for himself the law, but has actually constructed an elementary dynamo and motor for himself, together with instruments for measuring electrical currents. The law is then enunciated and further illustrated by a number of concrete examples, involving only the simplest mathematics, but at the same time practically covering every problem in electrical distribution by direct currents that the electrician is ever likely to be called upon to solve. The extreme simplicity of these operations—all depending upon Ohm's law—will doubtless be a revelation to most of those for whom this book is intended.

The dynamo and motor are treated as one and the same machine, as they really are, and it is shown that they all, of whatever type, are built upon exactly the same principles, and these are explained and developed from the simplest forms by the aid of numerous cuts and diagrams leading up to the finished multipolar machine of to-day. The intricacies and refinements of the completed machines, are, however, avoided as tending in a book of the scope of this one to undesirable confusion.

The theory and methods of speed control of railway motors are treated in as comprehensive manner as seemed desirable, the aim being always to explain the why and wherefore of a given arrangement, rather than to describe in full a specific device. Following this comes a brief description of some of the newer electric railway systems which promise to come into more or less general use, but which are not now generally known to the lay public.

Thus far it is believed that the book will be of service to others than motormen—that it will be of interest to that large class of intelligent Americans who desire to acquire a somewhat definite knowledge of this mysterious agent that they see working around them daily, who wish to know something more definite of electricity, either with the view of taking up its study more systematically in the future, or for the purpose solely of getting a more intelligent idea of its workings.

The last part of the work is devoted to directions for the care of street railway motors, the detection and remedy and prevention of their faults. These suggestions come from a most extended and varied practice, not of the author's alone, but of other engineers and of the manufacturers of the motors themselves. This portion of the book is especially intended for those whose business it is to handle the motors. The author hopes that this portion of the book will be found particularly useful to those for whom it is intended by reason of the fact that he has endeavored in the earlier chapters to so lead up to the subject that the *reason* for the various rules and precautions laid down will be at once apparent to the motorman, and that they will thus appeal to his intelligence much more strongly than they would if given empirically.

In conclusion the author wishes to acknowledge the assistance he has received by free reference to Crocker and Wheeler's "Practical Management of Dynamos and Motors," Crosby and Bell's "The Electric Railway," Silvanus Thompson's "Dynamo-Electric Machinery," and "The Electromagnet," and especially to the kind response to inquiries of The Westinghouse Electric and Manufacturing Co., The General Electric Co., The Short Electric Railway Co., The Sperry Electric Railway Co.

ELECTRIC RAILWAY MOTORS,

THEIR

CONSTRUCTION, OPERATION AND MAINTENANCE.

CHAPTER I.

INTRODUCTION.

Some years ago I became acquainted, as a boarder, with a bright young man who introduced himself to me as an electrician. It happened that we were placed at the same table, and as he was of an affable disposition we soon became very well acquainted. Whenever the subject of electricity came up, he was accustomed to speak in such a way as to impress all his hearers with the idea that he was an authority on all such subjects whose opinions were not to be questioned, and as he was generally correct in his statements I also became somewhat impressed with his knowledge, and would often ask questions—sometimes for information, and sometimes to see how nearly his opinions coincided with my own.

I found that he had been sent out by the then

leading company engaged in electrical railway construction, and his business in that city—a Western one—was the construction of its first electric railway. We soon became quite good friends, and when the power house was nearing completion I was a privileged character within its walls, for visitors were rigidly excluded as a rule; and later, when the first tests of the operation of the road were made, I was one of the fortunate few invited guests.

The first car over the line started out from the barns at ten o'clock at night, and carried as passengers, besides myself and the gentleman referred to, but three others—all officials of the road. Everything went smoothly until the further terminus—about three miles distant—was reached, when something gave way and we were stalled.

I must add here that my friend had not yet discovered that I was myself an electrician. I had had no other object in concealing the fact than that, as I had not been asked my business, it had never been necessary for me to declare it, and I felt that he would talk with me more freely if he did not know it. He knew that I was an engineer, however, and he attributed my ability to solve certain questions which arose to that fact—to that combined with what he took to be a smattering of electricity which he supposed I had acquired in some way or other.

As a matter of fact, however, I considered myself a full-fledged electrical engineer, and was, as electrical engineers went in those days; for besides having had a pretty thorough training on the purely theoretical side of electricity, I had worked in two of the largest shops in the country in all the departments of electrical manufacture. I had wound armatures and fields, and assisted in making and dressing down all the other parts of machines, both large and small; had assembled these various parts, connected them up, soldered the connections and finally tested the completed machines. I had also assisted in installing one road myself, and had inspected every road but one that was at that time in operation in the country, so that I was pretty familiar with electric railroad construction as then developed. I knew that my friend's experience had not been nearly as wide as my own, and I was glad to be with him on the occasion of the opening test of this, which I suspected was the first road over whose construction he had been completely in charge.

Well, as I say, something had happened, which was only really discovered when the attempt was made to run the car back. There was a considerable grade at starting, and this the car failed to ascend. We backed down to the level and tried it again. The car worked all right until the grade was reached, when she stopped again. This was repeated a number of times, and always with the same result. My own experience made me suspect at once the true trouble, but it would not do for me to suggest. My friend had been playing the rôle of a great authority on electric railroad matters for the special benefit of his invited guests on the car. He was in very high feather, and not without reason, that night, because the work of his hands was finished, and now he was showing his admiring guests how immeasurably better it worked than even his own anticipations had led him to hope.

For me to have suggested the cause would have been considered a piece of impertinence of the highest order, not only by him, but by the other guests, who looked upon my friend as a wonderful man; so I kept quiet.

My friend was a man of great self-confidence and equal to the emergency. In reply to inquiring glances, he said it was due to "something or other"—using a term not down either in Webster's or Houston's dictionary—that it was a very trivial matter and could easily be fixed; he would back down to the level again and fix it. He did so—that is, he backed down, and, after making an examination, said: "Yes, that's it, just as I thought; the 'something or other' was just what caused the trouble;" and he straightway proceeded to correct the thing. In the meantime I had had an opportunity of verifying my suspicions, but of course kept my counsel. When all was declared ready we entered the car again, and my friend, who really had done nothing, as far as I had been able to see—he certainly hadn't touched the root of the trouble—turning on the current to the last notch, said, "Now we go;" and we did until we had gotten even a little way beyond the point on the grade where we had always stalled before, and then we stalled again. We had gained a few feet over previous attempts simply because we had gained a little more momentum on the level. I think my friend understood this as well as I, but he pointed to the small gain with apparent pride, and with unblushing assurance stated that that proved that he was right in locating the trouble, and that now he would run back and fix the thing for good. We went back and he got under the car again. I watched him closely this time, and after doing absolutely nothing more than fumble around a little he came out and announced that "Now I've fixed her for good, and I'll show you how we can mount that grade."

I have often heard it claimed by horsemen, and have seen several instances myself where good results seemed to follow the plan, that to start a balky horse, the best way is simply to jump out of the wagon and pretend to fix the horse's bridle, then get in, and on giving the signal the horse, which neither whip nor oaths could budge before, would start right off as though nothing had happened. This was exactly what my friend seemed to have done and nothing more, and when, after we had all followed the injunction to get "aboard," and he commenced to turn on the current I could not but be astounded at the cheek, or rather the assurance, which enabled him to keep up the appearance of entire confidence in the success of another attempt which he *knew* must fail for *exactly* the same reasons the others had. It is needless to state what the result was, further than to say that we did not get quite as far this time as we had before, and it is also needless for me to state how, with equal composure, he gave some excuse for going back again to the starting point. I think he said he had forgotten his monkey-wrench or something else.

I determined this time to speak, and when we stopped at the bottom I got off and commenced looking around as if hunting for something, and stooping down, some distance from the car so as to get him alone, cried out, "I've found it," and, as I expected, he rushed over, leaving the others behind, and I whispered, "Look at your positive brush on the front motor." He heard me and understood. He stooped down, though, as if to examine what I pretended to have found, and then, in tones loud enough to be heard by the others, said, "No, that's not it, but I've got something that will do as well," and lost no time in getting under the car and repairing the broken connection which had rendered useless one of our motors.

When we got on to the car again and he grasped the controlling switch, I know I had more confidence that we could ascend the grade, but his appearance betrayed not one whit more than it had on the two previous attempts. As for the other parties, I am sure there was never a suspicion. We completed the return trip in great shape, and the trial trip was pronounced a success, and my friend was the hero of the hour. We returned to the car barns, and after he had seen everything safe for the night we walked home together. He seemed buried in deep thought for some distance, but finally broke the silence by, "You played me a darned mean trick." I was very much surprised at his attitude, thinking that he should rather have thanked me for helping him out of his difficulty. I knew that he would have found the difficulty after a while if left to his own resources, but thought I had saved him some embarrassment which further delay might have caused him, so I laughingly asked, "How?" to which he replied: "By not letting me know before that you were an electrician."

The above anecdote is related now merely to show how a little bravado, judiciously used, may save a reputation where entire candor would have been fatal. However reprehensible this practice may be in the abstract, it is one that pervades all walks in life, and is nowhere more prevalent, perhaps, than in the medical profession. It is, in fact, often used there to advantage, for if the physician should fail to inspire his patient with confidence in the beginning, it would be almost impossible for him to succeed later, and it is conceded that confidence in one's physician, as well as in the remedies one takes, is of the greatest help in curing disease.

If we find this practice so general in such an honorable profession as medicine, it is not surprising if we find it among the motormen, nor can we disparage it there while we uphold it elsewhere. In fact, I believe the motorman or electrical artisan is really less to blame than many others by the circumstances of his position. Most often the electric motorman has obtained his position as a reward for faithful services as a driver or conductor of a horse car. In his previous occupation his hours have been long and his work exhausting both to mind and body, and when his day's work is done he is in no condition for study or more work. Besides, as a driver, he has probably mastered his business, and there has been no incentive to further study.

With this habit of mind he is transferred without any special preparation to the responsibilities of the care of the electrical equipment of his car. He is broken in by someone who has had a little more experience than he has had, and after having been shown the various parts of his apparatus, and how they are intended to operate, is given a set of rules which tell him that he must do this and must not do that, and is allowed then to shift for himself. He is expected to talk about a current which he can't see, and to guard against results which he knows only by name. A new language is placed in his mouth which he does not understand, but which he understands somehow is intimately connected with his business. His hours are no shorter than before, and books which would explain matters are too expensive, even if he had time to read them, or entirely inaccessible. His companion motormen are using this new language familiarly, and he soon acquires that habit almost unconsciously, and as soon as he is thus initiated into the charmed circle which speaks this foreign language he becomes with his brother motormen a class distinct from that class from which they have all sprung. If he knows little of electricity, he still knows more than his former companions, and they treat him with a respect due to his superior knowledge. It will not do for him to admit ignorance, and he has an answer ready for every question. Nor does he care to show his ignorance among his companions by asking them questions or by asking questions of others in their presence, and in this way too often stands in his own light. That the average motorman is anxious to inform himself, when he can do so under circumstances which are not embarrassing, I know well from personal experience, for I have had individuals ply me with questions, when they could get me alone, who would rather have died almost than to have asked the same questions in the presence of fellow-workmen.

No one blames them for this. It is human nature, and, as I have stated before, they have shining examples in the same practice in members of professions considered more dignified than theirs.

CHAPTER II.

TECHNICAL TERMS, ETC.

WHENEVER we undertake the study of a new subject, it becomes necessary to, in a measure, break away from old things—old thoughts, old names, old tools, and oftentimes accustomed methods, and to adopt in their place new ones which those who have had experience in the subject have found more suited to the new. It is this first breaking away from the old and accustoming ourselves to the new that is the greatest stumbling-block in the path of the student. We can best describe a phenomenon or a thing to a person unfamiliar with it by likening it to something with which he *is* familiar, and then, after conveying to his mind a clear idea of how the new resembles the old, impress upon his mind as well as we can that the resemblance is not complete, and that the manner in which the familiar differs from the unfamiliar is really the essential difference between the two—we cannot at first explain this difference clearly; he must take that much on faith and assume at the outset that what is said is true. As the student becomes more and more familiar with the thing described, by handling it and seeing it used, he will begin to appreciate more fully its nature, and the peculiarities of the new object itself. It is, therefore, well in the beginning to substitute for the new thing an entirely new name instead of continuing to describe it by the names of things which it partly resembles, so that the mind may associate with the new name the actual qualities of the new thing rather than be misled by the name which the resemblance would indicate. This is true of all sciences, and it is true of electricity. Each branch of science presents some phenomena peculiar to itself, and therefore must have names for them, and it is the names peculiar to this science that constitute its technical terms. If a branch of science which we are about to take up is full of phenomena new to us, it will be full of strange names, and it is the mastery of these that is most formidable to us, for the meaning of each sentence is or may be obscured by their presence. It is frequently said that scientific men delight in obscuring their meaning behind technical terms, whereas the fact is they are using that language which is most intelligible to them, and is only unintelligible to us because of our ignorance of the subject.

If the student could only appreciate the fact that technical language is really the simplest and most direct that can be used, much of the difficulty of the subject at the outset would be removed, and it is therefore deemed wise to add a few words here with the hope of impressing upon his mind the truth of this statement.

In our everyday life we are constantly making use of technical terms as a matter of convenience; we do it unconsciously, it is true, but are impelled to it by the same necessities as is the electrician, the geologist or the astronomer. As an illustration, suppose we were to try to describe a dog to a person who had never seen one, or any animal of that family. We would have to describe it in some such vague terms as this: that it had four legs and a tail, was covered with hair, made a peculiar noise when angry or excited (imitating its bark), ate meat as its chief food, etc., etc, rehearsing many of its characteristics, which, however well described, would apply equally well to some other animal. Supposing, after having failed to convey in this way any adequate idea of the animal, we should secure a specimen and tell him that that was a dog; the story would be told a great deal better, and ever after the technical term "dog" would have a definite meaning to him—a much more definite meaning than any description that we could give. If his education stopped there—if he had seen but one breed of dog, and perhaps but a single specimen, all dogs to his mind would be about the same. If the one examined had been a great, shaggy-haired Newfoundland, he would not recognize in the hairless Mexican dog an animal of the same family at all; so that even after he had seen a dog of one breed it would be almost as difficult to describe to him another of a different breed as it was in the first place to tell him what kind of an animal a dog of any kind was. On the other hand, among dog fanciers how definite the phrase "Irish setter" is, for instance. Those two words—that technical phrase—convey to the mind more definite information than could be imparted in pages of printed matter, or perhaps more than in hours of discourse. To the person who was very familiar with dogs it would give an idea not only of the size and color, but of the general character, and yet how meaningless it would be to one who was not familiar with the term. To him it would not even convey the idea of a dog of any kind.

Thus it is that a technical term only has a meaning for us as we associate with it certain ideas. It seldom describes the thing itself, for it is impossible, as we have shown, to fully describe to another something he has never seen. The best we can do is to liken it to something that he has seen, and then caution him that the likeness is not exact. So that in describing electrical phenomena it must be understood that while our explanations and descriptions are the best we can give, they are not always exact, but only approximately true.

It will be apparent to everyone that there is more power available in a waterfall in which the volume or quantity of water flowing is great than in one where the quantity is small, and that the amount of power will be still greater if the water falls a great distance than if it falls but a short distance. In describing a waterfall, therefore, it is not sufficient to state either that it is of great height or that it is of great volume. We must state both the height and the volume. It is not sufficient to say that a waterfall is five hundred feet high, or that the water flows at the rate of one thousand gallons per second, but when it is said that a stream falls over a precipice five hundred feet high, at the rate of one thousand gallons per second, we know exactly how large the fall is and can figure out just how many horse power can be developed. Electricity is most often likened to a stream of water falling over a precipice, and the energy of an electric current is described in exactly the same way, only instead of measuring the height of its fall in feet, as we usually do water, electricians have decided to use the term "volt," and instead of measuring the flow in gallons per second they measure it in "amperes."

It is not necessary at this point to state just how much a volt or an ampere is—the terms are meaningless in themselves, and are the names of men who early did much to advance the science of electricity, that is all; but we must now try to *give* them a meaning. Using these terms in the above example, the waterfall would be described as falling five hundred volts at a rate of one thousand amperes. Perhaps we would better represent the volt as the equivalent of a pound of pressure, and then we can say of the water in a water pipe that it flows at the rate of so many amperes at a pressure of so many volts.

Now, for our purposes, we may consider the trolley and feed wires as copper pipes conveying water from a pump at the power station to a turbine or other water wheel beneath the car. The pressure of the water in this pipe is kept by the pump at five hundred pounds, and more or less gallons of it per second are used on the motor as we turn the controlling lever (faucet) on or off.

Everyone knows that if we twirl a wheel on an axle, be it ever so well greased, it will stop before long unless we continue applying power. It stops by reason of friction. Every carman also knows that, if his journals are not kept well greased, his car will pull harder and he will get a hot box. This is because there is *more* friction and the heat is generated faster than the rubbing parts can be cooled off, and therefore it accumulates to such an extent as to become not only apparent, but oftentimes troublesome. Now water, in flowing through a pipe, be it ever so smooth, encounters friction against the inside of the pipe. Heat is not observed in the case of water friction, because the water carries it away so rapidly, but the main effect is to retard the flow of water. Thus under a given pressure a pipe of a given size, say one hundred feet long, will deliver much more water per second than it will if the pipe were a mile or two long ; and again, more water will be delivered through a smooth pipe than through an equal length of a rough or rusty pipe, for the same reason that in the former there is less friction.

Now in the electric current we have pretty much the same state of affairs. Every conductor, however good, offers *some* resistance, in the way of friction, to the flow of current, and of two wires of the same diameter that will offer the greatest resistance which is longest. If one wire be twice as long as another, it will offer just twice the resistance, and that which is the smoothest inside, or in other words the best conductor, will offer the least resistance. Copper is the best conductor of electricity we have (except silver), and therefore our copper feeder wires and trolley wires may be likened to polished metallic pipes carrying water from our pump to our water wheel (motor) under the car, and the poorer conductors, such as iron or brass or zinc, may be likened to rusty pipes which produce more friction than the copper ones do. But the electric current cannot carry off heat in the same manner that water does, so, as in the case of the car axle, if the conductor be overworked it will get hot. Electricians have a way of measuring the resistance to the flow of current in conductors and express this resistance in "ohms." The word "ohm," like volt and ampere, has no meaning in itself, and is also the name of an early investigator, but to the electrician, who uses it to express the resistance due to friction, it has a definite meaning. Thus we have the three fundamental units of electricity : the volt, equivalent to a water pressure of say a pound to the square inch, or to a head of water of say one foot ; the ampere, equivalent to a rate of flow of water of so many gallons per second or minute ; and the ohm as the unit of resistance to a flow of water in a pipe, which for present purposes we may consider our conductors to be. We must bear in mind that as there are all sorts of pipes for conveying water—large ones, small ones, smooth and polished ones and rusty ones, all of which differ in the amount of water which they will deliver under a given pressure in a second or minute, according as they offer more or less frictional resistance to its flow—so are there different kinds of electrical pipes or conductors, which likewise differ in their carrying capacity of the electric current as they are large or small, smooth and polished (good conductors, silver, copper) or rusty ones (poor conductors, iron, brass, zinc, carbon, etc.); and although all of these units bear strange names —names of men who have distinguished themselves in scientific research—they are very closely equivalent to other units with which we have long been familiar.

CHAPTER III.

OHM'S LAW.

THESE three units bear a definite relation to each other also. Although common sense would seem to tell us that with a given pressure more water would be delivered through a short pipe in a given time than through a long one of the same diameter; that of two pipes of the same size and length that which was smooth inside would deliver more water in a given time than that which was rough or rusty or partly obstructed, and that of two pipes of the same kind, either smooth or rusty inside, that would deliver the greater quantity of water which was of the greater diameter—although, as I say, common sense would seem to tell us all this, electricians were a long time in finding out that it was true for electricity as well as for water.

It was George Simon Ohm who first discovered this simple relation of the flow of a current of electricity, but when he announced that the amount of current that would flow through any conductor was equal to the pressure (volts) divided by the frictional resistance (ohms), although he had really only stated that the electric current obeyed the same laws essentially as the flow of water through pipes, there were few who believed that it was true. Other investigators had imagined that the relation between the flow and the pressure and resistance was a much more complex one, and had gotten up long mathematical formulæ to express this supposed relation. Ohm's law, as it soon became called, was entirely too simple for them, and for a long time they would not accept it. Further experiment fortunately proved it to be strictly true, and this law, which is that the rate of flow of an electric current through a conductor, or the amperes, is equal to the pressure, or electromotive force or volts (all of which terms mean the same thing), divided by the resistance, or ohms, has become the very foundation of the science of electricity. It is usually written:

$$\text{Current} = \frac{\text{Electromotive force}}{\text{Resistance}},$$

or, for the sake of brevity, the initial letters are used only and the expression becomes:

$$C = \frac{E}{R}.$$

Nothing could be simpler than this when one understands it, and yet those who do not know any better imagine that the science of electricity is an exceedingly abstruse one. The contrary is really the fact, but unless one understands the A B C's he cannot read. It is worth while, therefore, that we devote some time in making Ohm's law entirely clear, for upon it is constructed practically the whole of the science with which we have to deal. The three letters

$$C = \frac{E}{R}$$

(amperes equals the volts divided by the resistance or ohms) constitutes the whole alphabet of our science, and we only need to know how to use these letters to solve any electrical problem with which we are confronted.

Let us illustrate the use of this law by a few numerical examples. Scientific men have very carefully determined the resistance in ohms of different sizes of pure copper wire, and have found that at the ordinary temperature a No. 10 B and S wire, which is a trifle over $\frac{1}{10}$ inch in diameter, offers a resistance of a trifle over 1 ohm per thousand feet. (The exact figures are $\frac{10189}{100000}$ inch in diameter and $1\frac{454}{10000}$ ohm resistance per thousand feet.) Let us discard the fractions and use the whole numbers. If we were designing a dynamo or buying one, we could obtain one that would give us any voltage we desired. In street railway practice a pressure of 500 volts or thereabouts is always used, and we will assume that we have a dynamo that will generate that pressure, and we have a circuit of No. 10 B and S wire of 10,000 feet, and we desire to know what rate of flow (how many amperes) can be sent over that circuit.

One thousand feet of No. 10 wire offer a resistance of 1 ohm; 10,000 feet will therefore offer a resistance of 10×1, or 10 ohms. The voltage or electromotive force at our disposal is 500. Ohm's law says that the current that will flow will be equal to the voltage or electromotive (in this case 500) divided by the resistance (in this case 10 ohms), and therefore the answer is that C (or amperes) $= \frac{500}{10}$, or 50 amperes. Now if we should make our circuit twice as long, or 20,000 feet, the resistance would be just twice as great ($20 \times 1 = 20$) and our equation would be $C = \frac{500}{20} = 25$. That is to say, with the same size wire of double the length the rate of flow of current would be only half as great, or 25 amperes instead of 50 amperes.

In the same way we find that if our circuit were only half as long, or 5000 feet, the resistance being also cut in two, the flow of current would be twice as great, or 100 amperes. Now if we double the amount of copper—that is, use two No. 10 wires instead of one—each of these will carry the same amount of current, and at 10,000 feet could deliver 2×50 amperes, or 100 amperes. That is to say, if we double the weight of our copper, either by using two wires of the same size or a single one of the same weight as the two combined, we will reduce our resistance by half, and consequently be enabled to deliver twice as much current at the same distance.

But supposing we have other data given. Suppose it is required to determine what sized wire to use to transmit say 1000 amperes to a distance of 20,000 feet—the electromotive force being, as before, 500 volts. Referring to Ohm's law

$$C = \frac{E}{R}$$

we have C, the amperes, equals 1000, and E, the pressure or electromotive force, is 500. Substituting these in the equation, it becomes

$$1000 = \frac{500}{R}.$$

By simple arithmetic this may be changed to

$$1000\ R = 500 \text{ and } R = \frac{500}{1000} = \tfrac{1}{2}.$$

Thus we find that the resistance of the conductor in ohms for 20,000 feet is $\tfrac{1}{2}$. The resistance of this same wire per 1000 feet will be $\frac{1}{20}$ of $\frac{1}{2}$, or $\frac{1}{40}$ ohm. A No. 10 wire has a resistance of 1 ohm per 1000 feet, and the required wire, which has a resistance of but $\frac{1}{40}$ ohm per 1000 feet, must be forty times as heavy as a No. 10 wire, or equivalent to 40 No. 10 wires, which by reference to any wiring table will be found to be equal to two 0000 wires.

Or supposing we have our wire already strung—say a No. 10 wire—and must deliver 1000 amperes over a circuit 20,000 feet in length, what electromotive force must we use?

The resistance of 1000 feet of No. 10 wire is 1 ohm. The resistance of 20,000 feet will be 20×1 ohms. The amperes or current which we have to deliver is 1000. By Ohm's law

$$C = \frac{E}{R}$$

Substituting for C its value 1000 and for R its value 20, our equation becomes

$$1000 = \frac{E}{20}.$$

or by simple arithmetic 20,000=E, which is to say that the electromotive force will have to be 20,000 volts to force a current of 1000 amperes around a circuit of 20,000 feet of No. 10 wire.

Examples have now been given of all possible cases of wire calculation in their simplest form, and these illustrate the way in which Ohm's law is used in electrical calculations. Of course others may and usually do arise in which some complications are introduced—for instance, it is usually required not simply to determine what is the smallest wire that can possibly carry a given number of amperes a certain distance, but to determine what size wire will carry that current the required distance with a given drop or loss of potential or electromotive force or pressure; but we will not discuss that question now, merely passing it by with the statement that it is a very simple matter to do, and requires no more knowledge of arithmetic than is involved in the examples already given.

It is strongly urged upon all who are not familiar with the use of Ohm's law, and who wish to derive benefit from the succeeding articles, to work out these problems over and over again until they are thoroughly familiar with the use of the law, for, as before stated, it is the key to the whole science of electricity. It will be observed that in Ohm's law there are three elements involved, viz., the pressure or electromotive force or voltage, whichever we choose to call it, the rate of flow of the current, or amperes, and the resistance which that flow encounters in the conductors, which is measured in ohms. These same three elements are involved in the flow of all other fluids as well, but are simply disguised under different names. As before stated, the flow of water in pipes involves pressure (usually measured in pounds per square inch or head in feet, corresponding to volts in electric currents), rate of flow (usually measured in gallons or barrels or cubic feet per minute or second, corresponding to amperes) and frictional resistance (usually measured in loss of feet or inches in head or in pounds pressure, corresponding to ohms in electricity), and, as shown by the examples, if any two of these three elements are given, the third may be determined.

RATE OF WORK.

In mechanics we say that when energy is expended at such a rate as to lift a weight of 1 pound 1 foot in 1 second it is doing work at the rate of 1 foot pound per second, and when it is doing an amount of work equivalent to raising 550 pounds 1 foot high per second, or, what is the same thing, raising 33,000 pounds 1 foot high per minute, the work done is equal to 1 horse power. A mechanical horse power is therefore defined as that expenditure of energy which will raise a weight of 550 pounds 1 foot high in 1 second, or 33,000 pounds (60×550) 1 foot high in 1 minute.

Since the raising of a weight usually conveys to our minds the idea of a lift or pull, rather than of a pressure, and we have heretofore been speaking only of pressures, it may be well, in order to make clear the exact similarity between the mechanical and electrical units of work to translate the "pull" or "lift" into a pressure. That "pull" and "pressure" are really equivalent will be apparent from a familiar example. When it is desired to move a car in the barns it is immaterial whether we get behind and push or get in front and pull. One method may be more convenient than the other, but the amount of work done in either case if the car be moved the same distance in the same time will be exactly the same, and if the work thus performed is the same as that required to lift or press upward a weight of 550 pounds through a height of 1 foot in 1 second it will be exactly a mechanical horse power. Thus we see that in the measure of rate of mechanical work which we call the horse power four elements are involved, viz., pressure, weight or quantity, dis-

RATE OF WORK.

tance and the time occupied in lifting or pushing the given weight or quantity through that distance.

Now in electrical language it will be remembered that the volt or electromotive force has been defined as the equivalent of mechanical pressure. We might also now say that it is the equivalent of mechanical pull or lift. We have defined the ampere as a rate of flow of current and likened it to the flow of so many gallons of water per second. While electricity is not supposed to have weight, we may for present purposes suppose that it has. A gallon of water weighs about 8 pounds, and if it is lifted through 1 foot in 1 second, 8 foot pounds of work will have been done. If we lift it or push it so fast that in 1 second we have lifted it through $\frac{550}{8}=68\frac{3}{4}$ feet in one second instead, we will have done work equivalent to 550 foot pounds in 1 second, which is equivalent to 1 horse power. Therefore, since the ampere is a similiar electrical expression to the mechanical expression of a flow of so many gallons per minute, and since the volt is equivalent to the mechanical expression of so many pounds pressure, and the product of pressure into gallons per second gives us foot pounds per second, 550 of which make a mechanical horse power, we ought to expect that the product of electrical pressure (the volts) into the rate of flow of the electrical currents (the ampere) would give us something similar to the foot pound per second, and that a certain number of these would be equivalent to a mechanical horse power, and so it is. If we have an electric current of 1 ampere flowing under a pressure of 1 volt, we have electrical energy expended at the rate of 1 *watt*, and this unit is of such a size that 746 of them are equivalent to 1 mechanical horse power. Or in other words, if 746 amperes under 1 volt pressure, or 1 ampere under 746 volts pressure, be wholly expended on an electric motor, that motor will be capable of lifting a weight of 550 pounds 1 foot high in 1 second, or 1 pound 550 feet high in the same time, or 33,000 pounds 1 foot high in a minute, or will be equivalent to a horse power. Thus while the watt is not exactly the same thing as a foot pound per second, it is a unit of exactly the same kind, but of a different size, just as although an ounce is not the same thing as a pound, it is a unit of the same kind. But 746 watts *are* exactly the same thing as a horse power, just as 16 ounces are the same thing as a pound.

Thus we have the electrical unit of rate of work or expenditure of energy also named after a distinguished early investigator, and it has no meaning whatever except that which electricians have assigned to it as described. The watt is equivalent to $\frac{1}{746}$ of a horse power, and the foot pound per second is $\frac{1}{550}$ of a horse power, so that the watt is somewhat smaller than the foot pound per second; but the electrical horse power and the mechanical horse power are exactly the same thing. The watt is also sometimes called the volt-ampere, because it is obtained by multiplying the volt by the ampere. The kilowatt is merely a thousand watts, the word *kilo* meaning thousand. Since 746 watts equal a horse power, a kilowatt is equal to $\frac{1000}{746}$, which is equal to nearly $1\frac{1}{3}$ horse power.

EXAMPLES.

How many electrical horse power can be delivered over a No. 10 wire whose length is 20,000 feet, the electromotive force being 500 volts?

Ans.—The resistance of No. 10 wire is 1 ohm

per 1000 feet. The resistance of 20,000 feet will therefore be 20 ohms. According to Ohm's law

$$C = \frac{E}{R}.$$

According to the problem the electromotive force or $E=500$, and the resistance or $R=20$. Substituting these in the equation, it becomes $C=\frac{500}{20}=25$. That is to say that 25 amperes can be delivered over 20,000 feet of No. 10 wire if the pressure is 500 volts.

The watts delivered will be equal to the electromotive force multiplied by the current in amperes, or $25 \times 500 = 125,000$ watts. Since 746 watts equal 1 horse power, the horse power delivered will be $\frac{125000}{746} = 167\frac{278}{746}$ horse power.

The current passing over a given circuit is 800 amperes, and the resistance of that circuit is known to be 15 ohms. What is the horse power expended?

Ans.—In this case $C=800$ and $R=15$. Substituting these values in Ohm's law the equation becomes

$$800 = \frac{E}{15}, \text{ or } E = 15 \times 800 = 12,000.$$

That is to say, the electromotive force on that circuit is 12,000 volts. The number of watts $= C \times E$ (amperes multiplied by volts), which in this case is $800 \times 12,000 = 9,600,000$ watts. Since 746 watts equal one horse power,

$$\frac{9,600,000}{746} = 12,868\frac{232}{746} \text{ horse power.}$$

The electromotive force of a given current is 500

volts, and the resistance of the circuit is 25 ohms. How many horse power can be delivered?

Ans.—$E=500$, $R=25$. Substituting these values, in Ohm's law the equation becomes

$$C = \frac{500}{25} = 20.$$

That is to say, the current that will flow under these conditions will be 20 amperes. Watts = $C \times E = 20 \times 500 = 10,000$ watts. (This may also be called 10 kilowatts.)

$$\text{Horse power} = \frac{\text{watts}}{746} = \frac{10,000}{746} = 13\tfrac{302}{746} \text{ horse power.}$$

Six hundred horse power are delivered over a given circuit, the electromotive force of which is 500 volts. What is the current?

Ans.—Since 1 horse power is equal to 746 watts, 600 horse power will be equal to $600 \times 746 = 447,600$ watts.

Since watts are equal to the volts multiplied by the amperes, there will be as many amperes as 500 is contained times in 447,600 watts:

$$\frac{447,600}{500} = 895\tfrac{1}{5} \text{ amperes.}$$

A street car motor is generating 10 horse power while taking 35 amperes. What must be the electromotive force of the current?

Ans.—1 H. P. = 746 watts. 10 H. P. = 10×746 watts = 7460 watts. Watts = $C \times E = 7460$. But $C = 35$. Substituting for C its value,

$$E = \frac{7460}{35} = 213\tfrac{5}{35} \text{ volts.}$$

With these examples the use of Ohm's law and the conversion of electrical into mechanical units and *vice versa* will have been sufficiently illustrated to show the extreme simplicity of the operation. It would be well for those desiring to really familiarize themselves with electrical problems to take figures which they may obtain from the actual operation of the roads with which they are connected and attempt their solution in the same manner. By watching the ammeter and voltmeter in the power house they can obtain an endless variety of problems as to how much work is being done on the line, and other data usually obtainable at the office will enable them to ring in changes on these problems which it will be not only interesting but exceedingly instructive to investigate.

CHAPTER IV.

THE ELECTRIC CURRENT AND ITS PROPERTIES.

WHILE all fluids resemble each other in some respects, each has some peculiarity of its own by which it is distinguished from every other fluid, and while we may partially describe one by showing its resemblances to other liquids or fluids, if we go no further than this, we will miss the very characteristics by which this particular fluid differs from the others. If we cannot liken these peculiarities to anything else, we can only become familiar with them by experimenting with the liquid itself and observing the peculiarities, and perhaps that will be the best way for us in this case. I think we can select a few experiments which will cost us little to perform and require little skill to prepare which will not only familiarize us with many of the peculiarities of the electric current, but will at the same time greatly assist us in understanding the principle upon which the electric motor operates, and it is proposed in this chapter to give a few of these, with full instructions how to prepare and perform them.

A word of advice here is that two or three motormen perform these experiments together and divide the expense of the material. This method will have several advantages, one of which is that where several work together it results in mutual benefit through discussions of the "whys" and "wherefores" of the phenomena, and this, it must be understood, is what we must strive after in all cases ; to know *why* a thing is so, as far as it is possible to know it, and in those cases where it is impossible to know *why* a thing is so we must endeavor to satisfy our minds whether, if it *is* so, it is *always* so or only occasionally or accidentally so. If it is *always* so, and we are satisfied of that fact, it is then a law. If it is only occasionally so, then there is surely some reason why it is ever so or why it is not always so, and we must not be satisfied until we have discovered this reason. That is the scientific way of doing things, and in fact the only satisfactory way. Another advantage in working together is that by sharing the expense of the material among two or three it need not exceed an amount which any motorman can spare. If there be three together, seventy-five cents apiece should buy everything that is needed, and I feel sure that the entertainment and benefit that will be derived from this investment will many times repay the outlay.

Thus far the flow of an electric current in a wire has been likened to the flow of water in a pipe, and in fact the resemblance is so strong that many electrical phenomena may be clearly predicted if we assume that our conductor is a small pipe and the electric current is a current of water flowing through that pipe urged along under a given pressure.

Some years ago the writer became acquainted with a German mechanic who was at that time in charge of the shops of one of our largest electrical factories, and who had already gotten out a number of inventions on minor details of electrical apparatus that were of the utmost benefit to the concern that employed him and which are in use to-day. In conversation with this mechanic one day he told the writer that he had never had any instruction in electricity whatever. Upon being asked how it was possible, then, for him to devise such excellent electrical apparatus, he replied that he had been educated as a hydraulic engineer and understood the mathematics of the flow of water through pipes thoroughly, and that when he wanted to know how the electric current would act under new conditions he simply assumed that he was dealing with water, figured out his problem accordingly, and was pretty sure to find that his results with electricity would correspond.

But while the flow of electricity corresponds closely in many respects with the flow of water, it differs radically from it in many others, and it is some of these differences that the following experiments are intended to show.

To start with, we must, of course, have a generator of electricity, and for this purpose any cheap primary battery will answer. If it is desired to make the battery one's self, it can be done very simply and cheaply and will prove both interesting and instructive. The galvanic battery depends upon the simple principle that if two unlike substances that are conductors of electricity are immersed at one end in a liquid capable of acting upon or corroding one of them more than the other, and the other ends of these (the two ends not immersed in the liquid) be connected together by a wire, a current of electricity will at once flow through that wire. Since almost all chemicals corrode metals to at least some extent, and hardly ever corrode any two metals to exactly the same extent, it would be almost impossible to select any two metals or any two conductors of electricity which would not form a battery if immersed in any solution we choose to employ. We therefore have an almost infinite variety of materials to choose from with which to make our battery, but of course they do not all make equally good batteries. If we can find two substances, one of which is very rapidly and easily corroded by our solution and the other not corroded at all, then we have the elements of a good battery. If the solution and both of the substances which are immersed in it are cheap, so much the better. Now it happens that zinc is both cheap and readily soluble in a large number of solutions, and carbon is also cheap and practically insoluble in all solutions, so zinc and carbon are usually employed in all modern batteries. It is a peculiar fact that a solution of table salt in water will not corrode zinc when acting upon it alone, nor will it corrode the zinc if there be placed in the same vessel with it a stick of carbon, so that the two might remain together in the same solution almost indefinitely, if they do not touch, without the zinc being corroded at

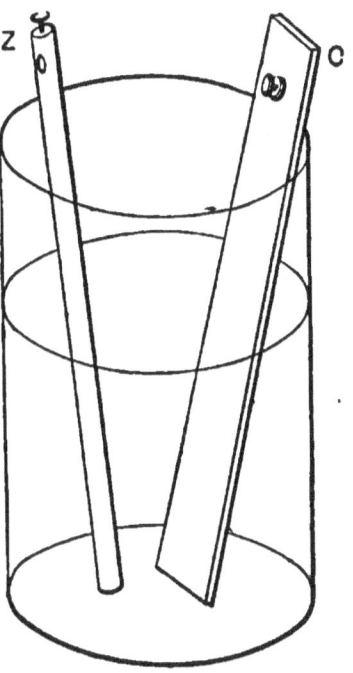

FIG. 1.

all, but if their outer ends be connected by a wire, the salt water will at once begin to attack the zinc, and an electric current will commence to flow, and continue until either the zinc is used up or the salt water has dissolved so much of it that it cannot dissolve any more—that is, providing the zinc and carbon remain connected by the wire, but the corrosion of the zinc will at once cease if this connection be broken. This property of salt water and zinc renders them pecularily suitable substances to employ in electrical batteries, because if the carbon and zinc be disconnected when the battery is not in use, there is no waste of material and the battery is always ready for use, requiring only that the zinc and carbon be connected again to start the current to flowing.

While an excellent battery may be made with a solution of ordinary table salt, there is still a better substance for this purpose, sal ammoniac, which is almost equally cheap and quite as harmless to handle.

To make a battery, procure a wide-mouthed jar either of earthenware or glass, and buy a stick of battery carbon about two inches wide, one-fourth inch thick and eight inches long, and also a "pencil zinc." These may be procured of any electrical supply dealer, and should both be provided at their upper ends with binding posts or means to facilitate the attachment of wires. The carbon will cost ten cents and the zinc five cents. At the same place or at the nearest drug store five cents' worth of sal ammoniac should be purchased, and we have all that is necessary to make a fairly good electric battery.

Empty the sal ammoniac into the jar and fill it three-quarters full of water, and when it is dissolved insert the carbon and zinc with their binding posts up. If we had an electric bell, and should connect one of its binding posts to the binding post of the zinc and the other to the binding post of the carbon, the bell would ring vigorously and continue to ring for a long time, showing that quite a strong current was flowing. If we had used table salt instead of sal ammoniac the only difference would have been that the battery would not last so long and the current would not be quite so strong.

Fig. 1 shows a battery such as described above, the zinc pencil being marked Z and the carbon marked C. In using such a battery, or "cell," as it is more properly called, it is better not to let the zinc and carbon touch each other in the liquid, and they *must not* be allowed to touch each other outside the liquid, for if they do it is the same as though they were connected by a wire, and if allowed to remain this way when not in use the battery will soon become entirely exhausted. For the reason that a little jarring might cause the zinc and carbon to come in contact with each other when the cell is not in use, it would be a wise precaution to remove either one or both from the liquid before setting the jar away.

It may perhaps be found more convenient to buy a battery, and if so a good dry battery, which can be bought for fifty cents, will be found as convenient as any. With the dry battery there is no danger of spilling any liquid, and the zinc and carbon are fastened in so that there is no possibility of their coming in contact with each other.

In a galvanic cell such as either of the above, the current is supposed to flow through the wire *from* the least soluble element to the one most readily corroded by the battery fluid. Where the elements are carbon and zinc as above, the current must always be considered as flowing *from* the carbon through the wire *to* the zinc. The carbon is therefore called the *positive* pole, and the zinc the *negative* pole, and correspond respectively to the positive brush of the dynamo, from which the current is supposed to flow, and the negative brush, through which it returns to the dynamo.

In addition to procuring the foregoing, an eight-ounce spool of No. 24 cotton-insulated copper wire should be purchased to complete the equipment for the following experiments. This can be had wherever the other supplies are purchased for forty cents.

In preparation for the experiments, cut out a piece of brown manilla or ordinary writing paper about two inches square, and wrap this carefully into a cylinder around an ordinary lead pencil. A round lead pencil is better for this purpose than an octagonal pencil. Next take the spool of insulated wire, and while the paper is still on the lead pencil, beginning at the left-hand end of the paper, overwind it tightly and closely with the wire until the right-hand end of the paper is reached. Then overwind this layer again with another layer, proceeding to the left, and then with a second layer winding to the right, and to prevent the wire from becoming loose at either end wind them with a few turns of strong thread or string and tie tightly. The wire may now be cut from the spool, leaving an end beyond the coil of three or four inches. There should be about the same length of loose wire left at the beginning of the coil. The paper cylinder with its overwrapped coils may now be slipped from the pencil and we have a hollow cylinder of paper overwound with three layers of insulated wire, the two ends of which, each some three or four inches long, extend out loosely, as in Fig. 2. The insulation should be carefully removed with a knife from about an inch of both ends of the wire to enable the coil to be placed in the circuit of the battery.

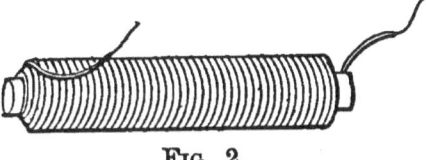

FIG. 2.

Cut off from the spool two more pieces of wire each about a foot long, and after removing the insulation from both ends of these for an inch or two, fasten one end of one in the zinc binding post and one end of the other in the carbon binding post. To the loose ends of the two wires twist the two loose ends of the small coil just made. If this is properly done, the carbon and zinc will be connected by wire, and the current will flow from the carbon around the various windings of the coil to the zinc. But as yet we have no evidence of this fact. Procure a large darning needle and insert it point first into the hollow coil, and after allowing it to remain there a few minutes take the needle out and examine it. No change will be observed—it is apparently exactly like the needle that was put in there a moment before, and yet one of the most remarkable changes known to science has quietly taken place. Cut a small piece of cork (about the size of a good-sized pea) large enough to float the needle in water, run the needle through this until the cork is in the middle and drop the needle with its float into a saucer full of water. The needle will swing around until it points exactly north and south. Reverse the position of the needle, or point it in any direction we choose, it will swing around so that it points north and south

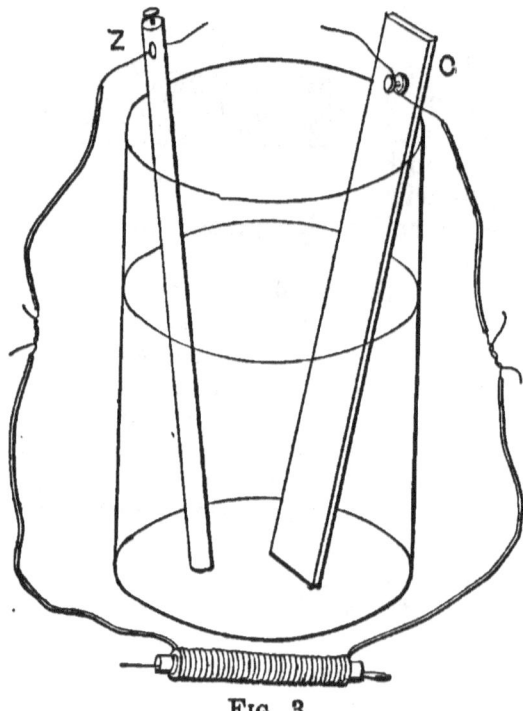

Fig. 3.

again. If on first trying, the point of the needle turns toward the north and the eye toward the south, it will always take up the same position again, the point always turning toward the north and the eye toward the south. The act of passing the current around the needle while inside the coil has given it this remarkable property, that ever afterward when free to move it will take up a north and south position, and the same end will always point north. We have magnetized the needle, and by rendering it free to move whichever way it likes have made one of the most remarkable instruments the world has ever seen, viz., a compass.

Place another needle in the coil in the same way—point first—and then float it on water. If the first needle took up a position with its point

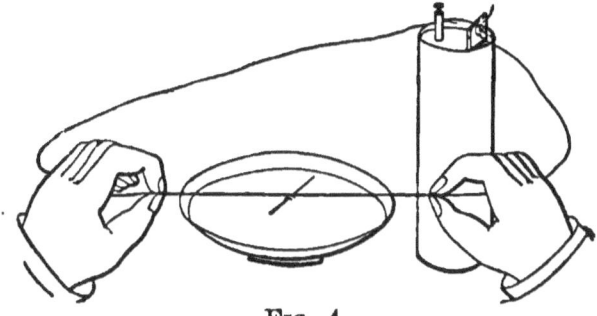

FIG. 4.

toward the north the second one will do the same, and when it has come to rest *its* point will be toward the north and its eye toward the south.

Try a third needle, but introduce this into the coil eye first. Float this and the eye will be toward the north instead of the point. We have discovered two laws, one, that a needle which has been placed in a coil through which an electric current is passing acquires the property of taking up a north and south position when free to move, the same end always turning toward the north; and second, that every needle which is placed in that coil in the same way will act in the same way, and the end which goes in first will always be the one to point north if that was the one that pointed north in the first experiment. For these reasons the north end of the needle is called the north-seeking pole and the other the south-seeking pole.

While one of the needles is floating in the water, if we bring the north-seeking pole of another needle near its north pole, it will rapidly repel it. If, on the contrary, we bring a north pole near a south pole, the two will attract each other strongly, and the floating needle will rush to and attach itself to the other so strongly that it may with care be lifted out of the water. In fact, we have two magnets with which we can produce all the phenomena of magnetism with which everyone is familiar.

Now let us remove our coil from the battery circuit and insert in its stead a straight piece of wire, say a foot long, by twisting its two ends to the battery wires, and then stretch it in a north and south direction directly over our floating needle or compass. It will be observed that the latter is deflected through a considerable angle from its former north and south position, and if the wire be *over* the needle the deflection will always be in the same direction. If we examine closely, we will find that if the current is flowing from north to south the needle will always be deflected to the east. If we reverse our wire, however, so that the current is flowing from south to north, the deflection will be to the west. Place the wire under the saucer, or even under the table upon which the saucer rests, and if the distance be not too great the needle will be again deflected, but in the opposite direction, viz., if the current flows through the wire from north to south *under* the needle, its north pole will be deflected to the west, and if the wire be reversed so that the current flows from south to north, the north end of the needle will be deflected to the east.

CHAPTER V.

THE ELECTRIC CURRENT AND ITS PROPERTIES.
(*Continued.*)

WE have now discovered one radical difference between the flow of water in pipes and the flow of an electric current in a conductor. In the case of water its flow within a pipe produces absolutely no external effect, but in the case of electricity we find that its flow produces quite a marked influence within the space surrounding the wire. In the case of the coil by which we magnetized the needles none of the current could possibly have gotten to the needles, first, because the wire of which the coil was composed was carefully insulated with cotton thread, which effectually prevented the escape of current from the wire, and second, there were several thicknesses of paper between the needle and the coil, and dry paper is one of the best electrical insulators known. We might have inclosed our needles in India rubber or glass before inserting them in the coil as a further protection against the current, but the result would have been the same exactly. We have also shown that if a wire in which a current is passing is held either above or below a compass needle the influence of the current upon the needle is manifested by a deflection either to the west or east of its normal north and south direction, and further, that we can not only thus tell whether a current is flowing in a wire or not, but we can tell in which direction it is flowing.

We have seen that if a current flows from north to south *over* the needle it deflects it to the east, and that it also deflects it to the east if it flows from south to north *under* the needle. If, therefore, we form a loop of our wire, in the center of which we place our saucer of water and needle so that the current flows first from north to south over and then from south to north under the needle (Fig. 5.), the effect of the one current is just doubled. If two loops are made, so that the current passes over the needle twice from north to south and twice under the needle from south to north, the effect of the current will be multiplied four times. By increasing the number of loops in this way we are enabled to produce quite perceptible deviations with such feeble currents that they could scarcely be detected in any other way. We have, in fact, actually made a *galvanometer*. As winding the wire above the needle in many turns would hide the needle so that we could not see its deflections, it is more usual in making galvanometers to wind the wire in a flat coil and place it under the needle. While the multiplying effect of a coil thus arranged is not as great as in the arrangement described, the instrument itself is much more convenient to use. To make such a coil cut out a piece of cigar box, or other lumber of

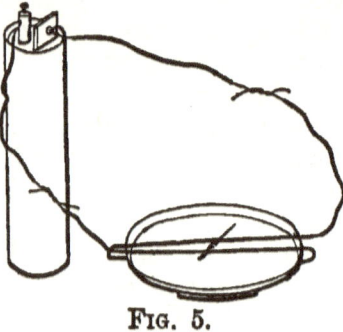

FIG. 5.

about that thickness, about 2 inches wide and 2½ long (Fig. 6). Cut away a seat on both ends for the wire and at opposite corners of one end bore a small hole about the diameter of a

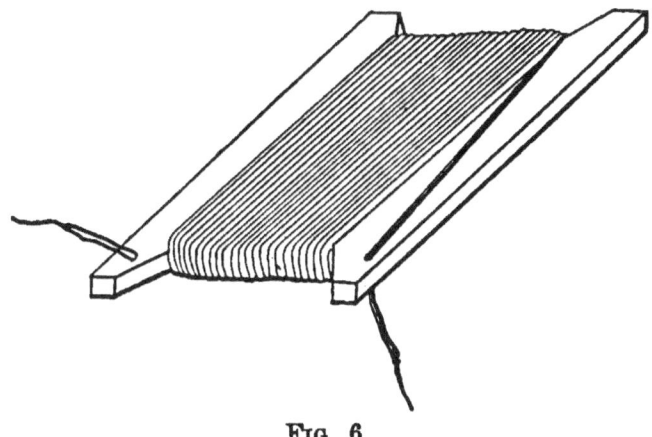

Fig. 6.

pin to hold the ends of the wire. Into one of these insert the end of the wire from the spool and pull it through about five or six inches and plug up the hole with a piece of wood so as to hold the wire tightly in place. Then wind on tightly eighty to a hundred turns of the wire and end it off through the other hole, plugging it up and leaving an end of about five or six inches as before. Now place the coil under the saucer and its floating needle and you have a very delicate galvanometer by which currents that would otherwise scarcely be suspected can be detected. Connect the two ends of the coil with the battery wires and the deviation of the needle will be found to be much greater and more rapid than in the previous experiments.

It must be borne in mind that the tendency of a current passing either above or below a compass needle is to place that needle at right angles to the direction of the current. The strongest current will therefore only cause the needle to take up a position at right angles to it—it will never cause it to deviate further or to reverse itself.

THE SOLENOID.

Now let us return to our first coil of wire (Fig. 2, see page 35). A hollow coil of wire such as this is called a *solenoid*, and it has some very peculiar properties which it will be necessary for us to understand before we can thoroughly comprehend either the dynamo or motor, and as we now have all the apparatus necessary to investigate these properties we will begin at once.

Take one of our magnetized needles and suspend it at its middle point by a very fine thread, so that the needle will hang horizontally and be free to move. Now connect up the solenoid with the battery and present first one end of the solenoid and then the other to, say, the north pole of the needle. It will be found that one end attracts and the other repels this pole just as would a real magnet, and that the end that repels the north pole attracts the south pole. In fact, although there is no iron or other magnetic material in the solenoid, it acts exactly like a magnet, and *is* one so long as the current flows through it. If the pole of the needle which the solenoid attracts be properly directed, the solenoid will suck the needle almost entirely within itself, and if the needle be reversed and pushed inside of the coil, it will be expelled again as soon as the fingers are removed from it so as to permit of it. Thus it

THE SOLENOID. 43

will be seen that a solenoid has a north and south pole, just as has a steel magnet, so long as a current is passing through it. Now detach one of the battery wires from its binding post, so that the current can no longer pass through the solenoid, and repeat these experiments. It will be found to be entirely inert, and will neither attract nor repel either pole of the needle. The magnetic property of the solenoid is therefore evidently entirely due to the current which is passing through it.

The phenomenon of sucking in or expelling a magnetized needle may be better illustrated, perhaps, by partly inserting the needle in the solenoid while one of the battery wires is disconnected, and then suddenly closing the circuit by touching the binding post with the disconnected wire. Thus far we have been experimenting with a highly tempered hard steel needle. We have found that when it is once magnetized it remains a magnet permanently. Now let us repeat our experiments, using a short piece (two or three inches long) of soft iron wire. Before we attempt to magnetize it let us try it with our compass needle. We find that it attracts both ends equally well. There is under no conditions any repulsion, because it is not magnetized. Place it in the solenoid as we did when magnetizing the needle, and while still in the solenoid test it with the compass. We find that one end attracts and the other repels either end of the compass needle just as did our permanently magnetized needle before, and even more strongly. Remove the wire from the solenoid and try it again. We find it repels neither end, but attracts both, as it did before it was magnetized. Suspend the wire in the middle and approach the solenoid to it. The latter will attract both ends equally well and suck in either end with equal facility. The soft iron wire is no longer a magnet. It was a stronger magnet while in the solenoid than the needle was, but immediately it is taken out or the current in the solenoid is broken it loses its magnetism. This may be more clearly illustrated by causing the wire, while in the solenoid, to pick up another small piece of the same wire, and then detaching one of the battery wires from its binding post. Immediately the current is broken the wire will drop its load, and it will remain incapable of picking it up again until the current is started in the solenoid.

There is, therefore, this very great difference between hard tempered steel and soft annealed iron, that the former when once magnetized retains its magnetism permanently, while the latter loses it immediately the encircling current is stopped.

There are also other minor differences, among which may be mentioned the following: Tempered steel requires an appreciable time to magnetize, whereas soft iron seems to assume the property instantaneously. It is somewhat difficult to change the direction of magnetism of steel—that is, after having magnetized a steel bar, so that one end is north and the other south, to demagnetize it and magnetize it again, so that the ends which were formerly north and south will become respectively south and north—while with soft iron the change is made with the greatest facility. A given current flowing through a solenoid of a given number of turns will make a much stronger magnet out of a soft iron bar than it will out of a steel bar of the same dimensions.

THE SOLENOID.

Within certain limits the amount of magnetism that can be imparted to a bar of iron or steel increases with the number of amperes of current passing through the solenoid and the number of turns in the coil. A limit is finally reached, however, beyond which neither an increase of current nor an increase in the number of turns or windings in the solenoid will materially increase the magnetism. With tempered steel this limit is much sooner reached than with soft annealed iron. For this reason, of two magnets of iron and steel of exactly the same size the one of soft iron can be made by far stronger than it is possible to make the one of steel. Cast iron, which partakes more of the nature of steel than of soft or wrought iron, is also inferior to the latter for magnetic purposes and loses its magnetism less readily. In fact, the best wrought iron that has been rolled or drawn, and not subsequently softened again by annealing processes, possesses the property of retaining some of its magnetism for quite a while and is less suitable for magnetic purposes, especially where it is desired to have the magnetism undergo rapid changes either of direction or intensity. All of these points have an important bearing upon motor and dynamo construction, and will be referred to again later on.

Another point which has also a bearing upon a subject to be taken up in a subsequent chapter may be referred to here.

As everyone knows, a knife or a piece of steel may be magnetized by rubbing it against another magnet. It may also be more slowly magnetized by laying the two side by side for some time even without touching each other. Now the earth itself is a huge magnet with a north and a south pole just like the small magnets we have made, and it is the attraction of the earth's north pole for the compass needle that causes the latter to point always to the north. Since two north or two south poles never attract each other, but repel, the end of the needle that points to the earth's north pole is really a *south* pole, and should not be called a north pole at all; this is the reason that when it was first referred to in this book, particular care was taken to call it the "*north-seeking pole.*" It is, however, usually called the north pole, because it points toward the north, and will hereafter be referred to by this name.

But to return to our subject. If the earth is really a magnet, we would expect bars of iron lying approximately parallel with the line joining its two poles to become in time magnetized, and as a matter of fact they do. Bars standing in a vertical position seem to become magnetized more rapidly, perhaps, than those lying horizontally, but those lying in a north and south direction, in clinging downward toward the earth at such an angle as to point directly toward the north pole, become magnetized still more quickly, and the magnetization is rendered almost instantaneous if the bar while held at this angle is smartly struck on either end with a hammer.

This experiment may be easily tried with an ordinary poker. After hitting it a tap on the end bring the end which was pointed toward the earth near the compass needle. It will smartly repel the north pole and attract the south pole. Reverse the poker and hit it another tap and test it with the compass; the polarity will be found to have been reversed. The end which at first repelled the north pole will now attract it and repel the south pole. Now there is a curious thing that we can do. We have literally knocked magnetism into the poker in the first place, and we have by reversing the poker and striking it again knocked a north pole out of one end into the other, and now we can knock the magnetism entirely out of the poker if we know how to do it. Hold the poker, which we will say is already magnetized, horizontally in an east and west direction and hit it one or two smart raps on either end. Now test it with the needle and it will be found to attract either end equally well, thus proving that the magnetism has entirely disappeared.

CHAPTER VI.

MEASURING THE CURRENT.

BEFORE passing on to another subject it may be well to call attention to something else we have accomplished in the simple instruments we have made. We have talked of amperes and volts and ohms, the three yardsticks by which we measure electrical phenomena, and have solved a number of problems involving these terms, but have said nothing, as yet, as to how the volts and amperes of a current or the ohms of a circuit are determined. We are all probably aware that the amperes are measured by an ammeter and the volts by a voltmeter, but we may not know on what principle either of these instruments is constructed. We do know, however, that if we have the volts and amperes we can determine the resistance in ohms by Ohm's law, but, as a matter of fact, we have constructed a device which, by slight modification, is capable of measuring both the volts and amperes.

We have already stated incidentally that the magnetizing power of a solenoid is equal to the number of amperes flowing through it multiplied by the number of convolutions or windings. To state it in another way, the amount of pull which a solenoid will exert upon a soft iron core partly inserted is also proportional to the product of the current in amperes multiplied by the number of turns. It may be well to prove this roughly, which we can do by suspending again our short piece of soft iron wire by a thread tied to its center and proceeding as follows: Connect up

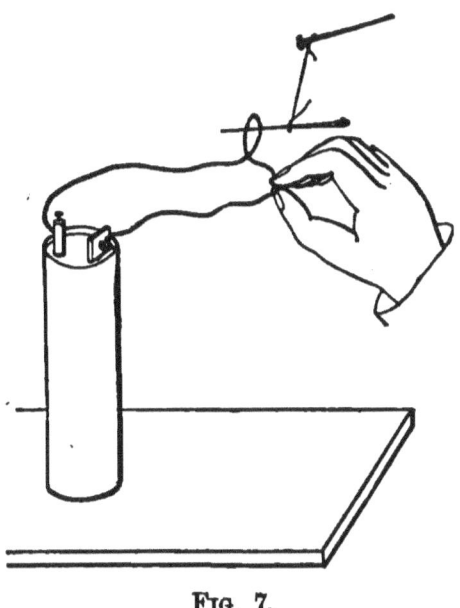

Fig. 7.

the battery with a few feet of wire and bend its center into a loop of one turn (Fig. 7) and present this loop to one end of the wire. We have here a solenoid of one turn, and it will tend to suck the wire into itself. The attraction will be comparatively feeble, to be sure, but much stronger than one would suppose who had not tried it ; but note as carefully as possible its strength. Next bend another loop so that we have a solenoid of two turns (Fig. 8). The pull will be perceptibly stronger. A third loop will increase the suction still more, and so on. If it were convenient for us to do it, we could show that if we could double the

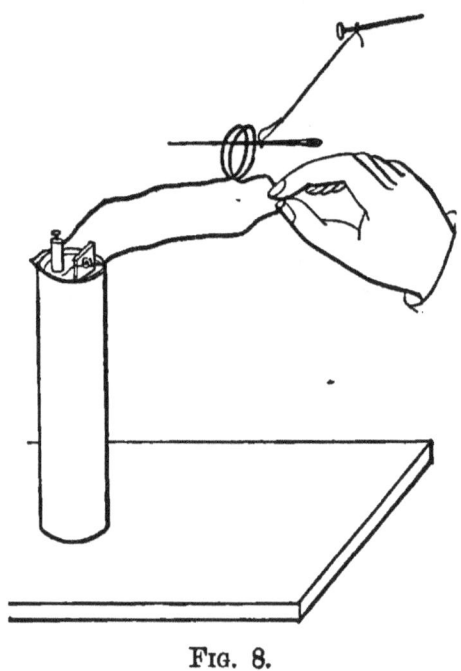

Fig. 8.

current in the single loop the increase of pull would be exactly the same as that obtained by adding a second loop. Now if we have a solenoid of any number of turns—it does not make any difference how many—and pass a known current through it, and measure the pull on the dial of a spring balance, or in any other convenient way, marking the point where the dial hand rests when the current is on full, and then measure the pull when twice this current and three times and four times this current are passing, marking each time the point where the dial hand rests, we will have a meter which will measure the current passing in terms of the unit used. If the first current used was one of one ampere and the second one of two amperes, etc., we will have an amperemeter or ammeter. Many of the ammeters used in street railway power stations are constructed after this plan, only the solenoid is made to lift a weight instead of pulling against a spring. Since the ammeter is intended to measure the full current, it is always placed in the circuit, so that all of the current passes through its coils just as we have placed it in our experiments thus far.

A voltmeter may be regarded as a more delicate instrument of the same kind, intended, however, to measure only a very small portion of the current. For this reason it is made of very high resistance, for, we know from Ohm's law, that of two circuits having the same electromotive force or voltage that will have the least current which has the highest resistance. Resistance alone would not be sufficient, however, for with the exceedingly small current used there would not be sufficient pull to operate the dial hand if the coil were made up of but few turns of a high resistance wire such as German silver. The coil is therefore made up of the best conducting copper wire, and the resistance is obtained by making this wire very long and bending it into a large number of loops or turns.

In operating an electric motor the latter does not consume amperes any more than a water wheel consumes or eats up gallons of water, but it does consume volts just as the water wheel consumes pounds of pressure. In the water wheel the water arrives at the wheel under a certain number of pounds pressure, and after it has gone through the wheel and done its work it flows away under a lessened pressure. The amount of pressure consumed by the wheel in doing its work will evidently be the difference of pressure at which the water arrives at the wheel and that under which it flows away. So with the electric motor if we wish to know the amount of electromotive force (volts) consumed in its operation, we measure the difference of potential at which the

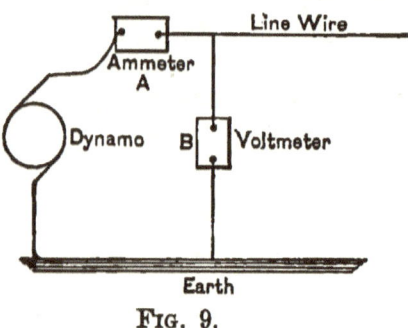

FIG. 9.

current arrives at and leaves the motor. The voltmeter is used to make this measurement, and is placed in a derived or shunt circuit, one end of which is connected with the main circuit at the positive side of the motor, and the other is connected with the same circuit at the negative side of the circuit, and the amount of current which will flow through the voltmeter and its circuit will be proportional to the difference of the potentials between the positive and negative sides of the motor. In Fig. 9 the method of placing the voltmeter and ammeter in circuit is shown.

From this it will be seen that all of the current which goes out to the car line passes through a few turns of the coil A, which, with its plunger and weighted lever, index hand and graduated dial, constitutes the ammeter, while a very small portion of the current is diverted from the outgoing wire to the return wire or the earth through the fine wire and coil of many convolutions, B, which, with its plunger, weighted lever, index hand and graduated dial, constitutes the voltmeter.

Of course there are many other kinds of voltmeters and ammeters constructed on nearly as many different principles, but we have not space nor is it worth while to describe them here. There is such an instrument as an ohmmeter, which reads directly the resistance of a circuit, but it is never used, so far as the writer knows, in street railway or electric lighting equipments. Resistances are more often measured by means of an instrument known as the Wheatstone bridge, by which the unknown resistances are compared with coils of known resistances, enough of the latter being added to just balance the unknown resistance, exactly as we weigh an unknown quantity of sugar or butter by placing the latter in one pan of our scales and adding pound and ounce weights to the other until the scales are exactly balanced. But as the motorman is not likely, as such, to ever have the handling of a bridge or be required to determine very accurately the resistances either of his circuits or his apparatus, and as it would only needlessly complicate matters at this time, its description will be omitted. But we already have the means of determining resistances with considerable accuracy in our voltmeter and ammeter, as before stated, for if we introduce these two instruments properly into our circuits and take their readings we learn the number of amperes flowing and the pressure or voltage under which that flow takes place, and by substituting these values in Ohm's law,

$$C = \frac{E}{R}$$

the resistances are readily calculated.

MAGNETISM AND ELECTROMAGNETISM.

In discussing the electric current we likened it to a flow of water or other fluid through pipe, and showed that this flow is retarded in both cases by obstructions, and that in both cases these obstructions produced like results, which might be predicted by the simple relation between the flow, pressure and resistance expressed in Ohm's law. We further found that in the case of electric current the influence of the flow within the wire extended to the space surrounding the wire, and that the effect of this influence was to produce magnetism in that space, or, as electricians would say, to produce a magnetic field. Thus every conductor carrying a current of electricity produces in the space surrounding it a magnetic field. That this is true our experiments first showed by the magnetizing effect which the solenoid had upon the needle placed within it, and further, by the tendency of the floating magnetic needle to take up a position at right angles to a current passing in a north or south direction either above or below it. That the magnetic field thus generated by the current has a definite north and south pole, which will be later shown is dependent upon the direction in which the current flows in the wire, was proved by the fact that one end of the solenoid repelled one end of the needle, while it attracted the other, just as did another magnetized needle—this being the test of magnetic polarity.

It has also been shown that a piece of hard tempered steel, and to a less extent cast iron and even wrought iron which has become somewhat hardened either through drawing or forging or too rapid cooling from a high temperature, retained the magnetism indefinitely which had been imparted to it by the magnetizing current, but that the magnetism of the solenoid, as well as that of soft annealed iron surrounded by a solenoid, lost its magnetism the moment the flow of current ceased. These two phenomena give rise to two classes of magnets, viz., permanent and electromagnets, which differ from each other only in the fact that in one case magnetism when once produced continues practically unchanged after the magnetizing current has been withdrawn, and in the other it is wholly dependent upon the continuance of the current for its existence, and the strength of the magnetism varies with every instantaneous change of the strength of the current.

Since magnetism is produced whenever an electric current flows through a conductor, and in the case of the electromagnet varies instantaneously with the variations in that flow, it is readily suggested that the two, electricity and magnetism, are closely related and may follow somewhat similar laws, and experiment has demonstrated the correctness of this idea.

While from the fact that a permanent magnet continues to be magnetic even after the magnetizing force is withdrawn we do not associate the same idea of *flow* in a magnet that we do in the case of an electric current, and while there really may be no flow or motion of the thing which we may for the want of a better name call the "magnetic fluid," still the greatest advance that has ever been made in the science of electricity (of which magnetism is a most important part) was due to this conception that an actual flow of force does take place in the magnet, its direction being from the north pole through the air to the south pole, and thence through the magnet back again to the north pole. This idea involved the idea of a magnetic circuit similar to that of the electric circuit, which must contain resistances. And, as in the electric circuit, in order to maintain the flow of current there must be some pressure to force the fluid through these resistances. It would logically follow from these assumptions that if a certain flow could be forced to take place against a given resistance by a given pressure, a greater flow could be maintained through the same resistance by a greater pressure, and we would have for magnetism another Ohm's law expressing the relation between the amount of flow (strength of magnetism), the pressure, and the resistance offered to the flow. Experiment has fully verified this hypothesis, and we have the law of magnetism that the strength of a magnet is equal to the pressure divided by the resistance of the circuit. While the pressure in the flow of water is usually spoken of as hydrostatic pressure, and that in the flow of electricity is called *electro*motive force, the pressure which causes magnetism in a magnetic circuit is called *magneto*motive force.

It has already been stated, and the experiments illustrated in Figs. 7 and 8 (pp. 49 and 50) have shown, that the magnetic field produced by a current flowing through a solenoid consisting of two turns of a wire is twice as strong as that produced by its flowing around a solenoid consisting of but one turn. It has also been stated, and it is equally true, that the magnetizing effect of a coil of wire of any number of turns will be doubled if the current in amperes flowing through the wire is doubled, and it will be three times as strong if the current is increased threefold. The law of the magnetizing effect of a solenoid or hollow coil of wire may therefore be stated as follows : It is proportional to the current flowing and to the number of times it flows through or around the space which constitutes the magnetic field. Or in other words, it is proportional to the product of the amperes multiplied by the number of turns or convolutions in the solenoid. Since an electric current is always measured in amperes, and a coil of wire may be definitely described by the number of turns of which it is composed, it is a convenient way of expressing the magnetizing effect of the combination of the two by the product of these two, and the magnetic effect produced by a current of one ampere flowing around a coil consisting of a single turn of wire, usually called an "ampere turn," has been adopted as the unit of magnetomotive force. Thus the magnetomotive force of a solenoid may be said to be that of 100 ampere turns. It matters not how these 100 ampere turns are made up—whether a current of 1 ampere flows through a coil of 100 turns, 2 amperes through 50 turns, 25 amperes through 4 turns or 100 amperes flow through a coil of a single turn—provided only that the product of the amperes and the number of turns in the coil is equal to 100, the magnetomotive force will always be the same.

But we have seen that the quantity of electricity (amperes) that will flow through a circuit is not alone dependent upon the electromotive force, but is also dependent upon the resistance of the circuit, being greater where that circuit is short or has little resistance, and less where it is long or has greater resistance. Ohm's laws says that it is always equal to the electromotive force divided by the resistance. This is equally true of magnetism. When we have expressed the magnetomotive force of a magnet, we have not yet defined its strength, because we have not mentioned the resistance of the circuit through which the flow is assumed to take place.

CHAPTER VII.

LINES OF FORCE.

WE have the greatest possible variety of conductors of electricity, from silver and copper, which conduct it with the greatest facility and offer the least resistance, down through all of the other metals and carbon to substances which are non-metallic in character which offer very great resistances, and are usually termed "non-conductors." But of conductors of magnetism—that is, magnetic substances—we have but three: iron, nickel, and cobalt, of which iron possesses the property in a pre-eminent degree. All other substances, including the air, stand about upon a par with each other; hence it is that a magnet acts quite as well through the top of a table, through a teacup or saucer full of water, as it does through the same space of air.

If, therefore, we increase the air gap between the north and south poles of a magnet, we increase the resistance to the flow of magnetism, for the latter has to traverse this length of non-conducting or poorly conducting substance, and with an increased air space it will, as in the case of an increase in the resistance of an electric circuit, require a corresponding increase of magneto-motive force—or, what is equivalent, an increase in the number of ampere turns of wire—to produce through this increased resistance an equal flow of magnetic fluid.

With the conception of a flow of magnetism, or "flux," as it is technically called, it became necessary to have some unit corresponding to the ampere as used in the flow of electricity. For the magnetic flow or flux the unit adopted is "the

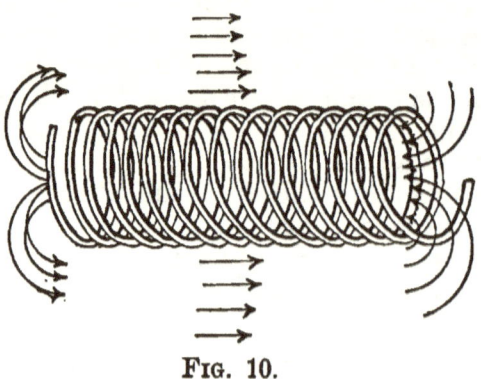

FIG. 10.

line of force," which is an imaginary line passing out of the magnet at the north pole and into it at the south pole.

Thus in Fig. 10, which represents a solenoid, the lines of force emerge at the north pole and pass around through the air and re-enter at the south pole. In the case of the solenoid the whole of the magnetic circuit is through the air. The resistance to the flux is therefore very great, and the number of lines of force, or, in other words, the strength of the magnet, will therefore be very small.

In an electric circuit, if we substitute a good conductor for a poor one, we will get, with the same electromotive force, more amperes, so if in the solenoid (Fig. 10) we substitute a bar of iron, a good conductor, for the air inside of the coil, we will have greatly reduced the total resistance of the circuit, and the number of lines of force that will flow around the magnetic circuit will be enormously increased, which means that we have

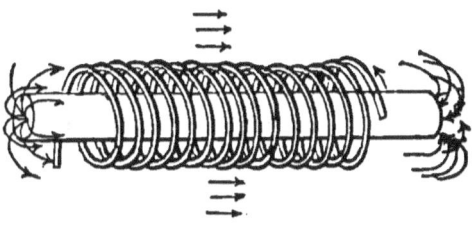

Fig. 11.

a magnet of enormously greater strength from the same number of ampere turns.

We have here (Fig. 11) a bar electromagnet. But strong as this is it is evident that in any bar magnet at least one-half the magnetic circuit must be in air, and the longer the bar the longer will be the portion of the circuit which must traverse the high-resisting air. If we bend the bar around in the shape of a horseshoe, bringing the two ends near together, we may greatly reduce the distance which the lines of force must traverse the air, and this means a still greater number of lines that will be caused to flow by the same number of ampere turns. Thus in Fig. 12 the N. and S. poles are brought close together, so that the air space between them is very short. The lines emanating from the face of the N. pole and entering the face of the S. pole will therefore be very dense—much denser than was the case in Fig. 11, where the iron bar was the same length and was excited by the same number of ampere turns.

In making these sketches to any given scale there is, of course, a limit to the number of lines of force that we can draw. We can draw a certain number, but there is no room for more.

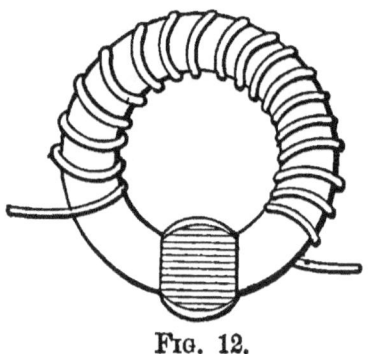

Fig. 12.

The same conditions exactly exist as regards the number of actual lines of force that can be made to thread through any given bar of iron. Up to a certain point the number of lines remains very closely proportional to the magnetizing force (ampere turns) divided by the magnetic resistance, but a limit is finally reached beyond which the increase of ampere turns has very little additional effect upon the strength of magnetism produced—new lines of force are not added, simply for the same reason that we cannot draw any more in our sketch, viz., that there is no room for them. When this condition is reached, the iron is said to be "saturated."

We have heretofore spoken of these lines of force as purely imaginary lines. They have, however, a more real existence than this, for they can be traced or made visible in various ways. If a plate of glass be sprinkled with fine iron filings, and then held close to a magnet and be gently tapped so as to allow the particles of iron to take up any position they desire, they will arrange themselves in curved lines emanating from the north pole and re-entering the south pole, which correspond in length and direction with, and, in fact, are, the visible representatives of what we have been speaking of. They may also be traced with a small compass needle, which, in whatever position it may be held with reference to the magnet, will have exactly the same direction as the line of force which passes through it has at that place. Thus a compass neeple points in a north and south direction, simply because all lines of force passing between the two poles of the earth have a north and south direction. If, however, we bring another magnet near the needle, the lines of force emanating from it will overpower those of the earth's magnetic poles, and the needle will take up a position in accordance with the stronger, or rather in a position which is a compromise between the two.

THE CLOSED MAGNETIC CIRCUIT.

By reference to the lines of force shown by the iron filings it will be seen that they emerge from and re-enter the magnet only at the poles. We may therefore define the magnetic poles as those portions of the magnet where the lines of force emerge from and re-enter the magnet. We have also seen that as we decrease the air distance through which these lines have to pass, by bending the magnet around so that the poles come closer together, the magnet increases in strength. It would be logical, therefore, to assume that if we brought the two poles into actual contact, or, in fact, made a closed ring of the magnet, we would have a magnet of maximum strength, because the air resistance would be reduced to nothing, and such is really the case, and the smaller the diameter of the ring the stronger will be the magnet, because the lines of force will have a less distance of iron to traverse, and hence meet with less resistance due to that cause.

But according to our last definition a magnetic pole is that surface whence the lines of force either emerge or where they enter the magnet. Since iron is such an enormously better conductor of magnetism than the air, the lines of force will continue in the iron even though they may have to go considerably out of the shortest path to do so. There will in this case be no surfaces from which the lines will emerge or into which they will enter, and therefore there will be no poles. If there are no lines of force external to the magnet, there will be none to direct a compass needle, and the latter will not be deflected as by an ordinary magnet, nor will this ring magnet attract other particles of iron or steel. This is entirely contrary to the popular conception of a magnet, and it is only in the more modern works that a magnet is not defined by the unqualified statement that it possesses both north and south poles which have the property of attracting to themselves other magnetic substances. We have, however, learned that we obtain the strongest magnet with the least expenditure of energy in an absolutely closed magnetic circuit and by making that circuit as short as possible, and yet this strongest magnet has neither north pole nor south pole, nor will it attract or repel other pieces of iron or steel.

MAGNETIC LEAKAGE.

It is seldom, however, that we can realize fully these ideal conditions. Although our iron ring may be of the softest iron, and there may be plenty of it to carry all the lines of force that are generated by the ampere turns used, some of these lines are apt to wander outside the iron. and to take a short cut across the air space. These may be readily detected and their direction indicated by the deviation of the compass needle. To such straying lines the term " leakage lines " or " magnetic leakage " has been given.

While the unbroken ring or closed magnetic circuit gives us by far the strongest magnet for the material and energy employed, there are many uses of the magnet which require that it should have polarity—that its lines of force should pass, for a portion of the distance at least, through an interval into which may be introduced substances to be acted upon by these lines ; but it is a cardinal law of magnets that that magnet which most nearly approaches the closed magnetic circuit will be the most efficient. For this reason magnets, whether permanent or electro magnets, are usually bent around so that their poles approach each other, and the object to be magnetized is introduced into the gap between the two poles. In order to concentrate the lines and make them as dense as possible in this gap it is usual to wind the substance to be acted upon on a core of iron, or imbed it in its mass, and this is introduced into the gap, reducing by that much the air resistance. Of course all leakage lines, or those which do not pass through the path intercepted, are wasted, and whatever of current was required to generate the leakage lines was uselessly expended. Exactly parallel would be the case were we pumping water through a long pipe to operate at its other end a small water wheel. If the pipe were full of small holes along its length, just as much more water would have to be pumped into the pipe to do the same amount of work as leaked out through these holes. That is to say, the water that leaks out will cost us just as much to pump, per gallon, as that which issues from the nozzle, and yet it does no useful work. The economical man will therefore not use a sieve for a water pipe, and the builder of magnets will avoid shapes which tend to magnetic leakage. In designing magnets it is always desirable to keep the two sides—the north half and the south half—as far away from each other as possible except at the poles, so that the air gap between the latter where we want to utilize the lines of force will be the path of least resistance, and the great distance between all other portions of the magnet of different polarity will offer too great a resistance for lines of force to jump across.

Sharp angles or points should also be generally avoided, because magnetism leaks more readily from a point or angle than from a smooth surface. A magnet of a circular ring shape best meets the required conditions, although a strict adherence to that form is not always practicable or even desirable.

CHAPTER VIII.

POLARITY, MAGNETISM AND CURRENT.

For the purpose of showing the lines of force I bought for ten cents a small horseshoe magnet $2\frac{1}{2}$ inches long. Laying this on its side on a table, I placed over it a piece of glass which had been varnished on one side and allowed to become perfectly dry. Upon this I sifted as evenly as possible some fine iron filings, and then tapped the plate gently on the edges in order to permit the filings to arrange themselves. The beautiful curves shown in Fig. 13 were the result. The glass was then carefully lifted from the magnet and heated over a gas flame. This softened the varnish so that the filings were stuck to the plate, and when the varnish had hardened again were preserved for the making of this cut. Fig. 13 shows the lines of force emanating from this magnet when the armature or keeper is entirely removed. Fig. 14 shows the lines as they were when the armature was removed about one-third of an inch from the poles, and Fig. 15 shows the lines as they appeared when the polar ends of the magnet alone were presented to the under side of the glass plate.

We may look upon these lines of force as so many elastic bands or strings by which the magnet attaches itself to other pieces of iron.

When the iron is in actual contact with the poles of the magnet, it is bound to the latter by a great number of these strings. When we endeavor to pull the iron away, we have to pull against the

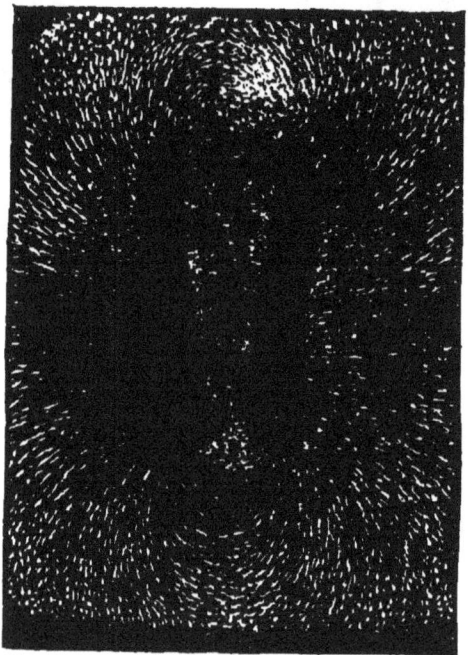

Fig. 13.

combined elasticity of all these strings. The moment we succeed in pulling it the slightest distance from the magnet a great many of these strings snap or pull out of the iron and disappear in the magnet just as india-rubber strings would if they came out of a hollow tube and were attached to the piece of iron we were trying to pull away. As we remove the iron still farther more and more of

POLARITY, MAGNETISM AND CURRENT. 69

these elastic strings or bands snap, until the iron is removed beyond the attraction of the magnet, when it may be said that all of the bands have been snapped.

Reversing the operation by gradually approach-

Fig. 14.

ing a piece of iron to the poles, we will have to draw upon our imagination somewhat for an equally good illustration. As it comes within the attraction of the magnet first one or two elastic bands jump out of the north pole, thread their way through the iron and attach themselves to the south pole. With a nearer approach a great many more do the same thing and pull the iron to the poles with all the force of their elasticity, and finally as the iron comes nearer they come out with a rush, and with their combined pull hold

Fig. 15.

the iron to the poles with the greatest force of which that particular magnet is capable.

Now a clear conception of the behavior of these lines of force is indispensable to an intelligent understanding of the theory of the dynamo and the motor, and it is for this reason that so much space has been devoted to the subject. Do not be deceived with the idea that these lines of force are solely of theoretical interest. We discussed them at first as though their existence was purely imaginary. Then we showed by means of the iron filings and the photographs that they really do exist. Next we discussed their behavior in a somewhat theoretical manner, and

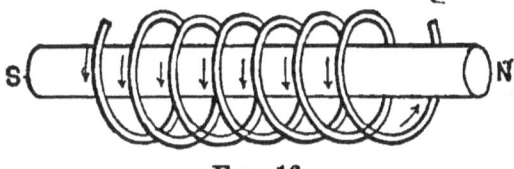

FIG. 16.

shortly we will show how upon this behavior depends entirely the action of both the dynamo electric machine and the electric motor. But before taking up this latter task, which, to make easily intelligible, so much has been said by way of preparation, a few words must be added as to the effect which the direction of flow of the current in the solenoid has upon the polarity of the resulting magnet.

POLARITY.

It may be stated at once that the way in which the wire of a solenoid is wound has absolutely nothing to do with which end of the inclosed iron bar will be the north pole and which the south, but all depends upon the direction in which the current flows around it.

If in looking at the end of a coil the current flows around it in the direction of the hands of a clock, that end will be a south pole (Fig. 16). If the current is flowing in the opposite direction to that pursued by the hands of a clock, or from right over the magnet to left, that end will be a north pole (Fig. 17). It is therefore merely a matter of convenience whether we wind a magnet right-handed or left-handed; we can make either end a north pole and the other a south pole by simply connecting the ends of the coil with our circuit so that the current will flow around the magnet in the proper direction, and if at any time we wish

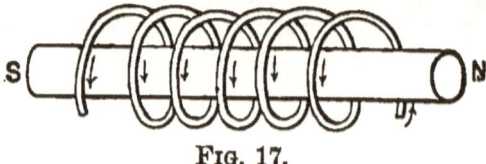

FIG. 17.

to reverse the poles of the magnet we simply have to reverse our connections. It is well to bear this in mind, for there is a popular fallacy that the direction of winding the magnet, right-handed or left-handed, determines which end will be north and which south, but as a matter of fact it has nothing to do with it.

MAGNETISM AND CURRENT.

For the purpose of illustrating the part which the lines of force (magnetism) play in the generation of current I went to a blacksmith's shop and cut off a piece of ⅜-inch round iron 5½ inches long. This I heated and bent into the shape of a hairpin (*A*, Fig. 18), bringing the ends to within 1¼ inch of each other. I then cut off another piece from the same rod and bent its two ends upward at right angles (*B*, Fig. 18), so that the distance between these two ends was the same as between the ends of *A*. The ends of both *A* and *B* were then filed smooth, so that when placed together the surfaces rested flatly against each other. Any blacksmith would probably have done this job for me for twenty-five cents.

I next made two paper spools, each an inch long, by wrapping several thicknesses of manilla paper around a stick whittled to about the same diameter as my iron, and forcing onto this paper cylinder two washers made of heavy cardboard, and pasting the upturned ends of the paper cylinder to the outside surfaces of these washers, so as to hold them in place. Then, while the spools were still on the stick, I wound on them tightly and as evenly as possible layer after layer of insulated wire until the spools were full. I counted the number of turns on the first spool, and it happened to be 382 turns. I wished both coils to be as nearly alike as possible, so in winding the second spool I put on just the same number of turns. Then, slipping the spools off of the stick, I slipped one on each leg of my bent iron rod A until about a quarter of an inch or less of the ends protruded; one end of each of the coils, having been cleaned of its insulation, were twisted together and the other ends were connected to the battery. The connections of the coils with each other were such that, in whichever way the current passed, when we looked at the two poles it would be clockwise around one and counter clockwise around the other. Immediately the current began to flow in the coils

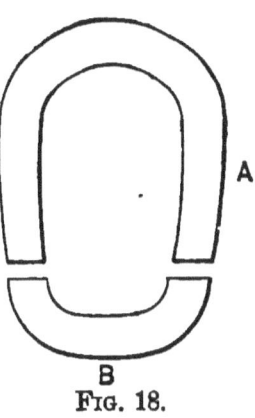

FIG. 18.

the hairpin, or bent iron A, became a most powerful magnet, and upon bringing the ends of B in contact with its poles it attracted it with great force. Although the iron A weighed less than a quarter of a pound, it sustained a weight of about ten pounds attached to the armature B. That such a small amount of iron could support such a weight would seem almost incredible to one who had not witnessed it, and, in fact, would have been utterly impossible with a permanent magnet of the same weight. On breaking the current the magnetism disappeared at once, as has already been explained. It was found, however, that it would still hold up a steel penpoint or two, showing that its magnetism was not entirely lost. This *residual* magnetism, as it is called, was probably due to the fact that the iron had become somewhat hardened either by the hammering it was subjected to on the anvil or by too rapid cooling after it came from the forge. After it had been left for twenty-four hours without current, however, the residual magnetism had become so small that it would no longer support even the smallest piece of iron accessible, although slight traces of polarity were detectable on presenting the two ends successively to one end of the floating needle. These tests, therefore, showed that it fairly answered all the requirements of a good electromagnet.

CHAPTER IX.

ELECTROMAGNETIC INDUCTION.

I NEXT constructed a paper spool on the armature B, similar to those placed on the legs of A, and wound it as full as I could get it of insulated wire. I counted the number of turns and found it to be 421. I may say here that all of these dimensions were accidental, and are only given to show what actual results were obtainable from them.

The object of placing the coil on the short piece B, which we will hereafter designate as the armature, was to show what effect the snapping of the lines of force which thread a solenoid, or their sudden appearance in the same, would have upon the solenoid. It is evident that if no current be traversing the coils on A there will be no lines of force traveling around them. If, however, we place B in contact with A, we have practically a closed magnetic circuit, and if, after having detached one of the wires from the battery, we touch it to its binding post, the full strength of the magnet will be instantaneously developed, all of the lines of force of which our apparatus is capable will be instantaneously developed in A, and will rush out of the north pole, thread their way through the armature B, which is the core of our solenoid, and enter the magnet again at the south pole.

If we now break our electric current again, the lines of force traversing the magnet and the armature coil will be as suddenly snapped and disappear, the result being the same, though more effective, perhaps, as if the armature *B* were suddenly jerked away from the magnet while the latter retained its full magnetic properties. To discover the effect of the sudden introduction within or withdrawal from the armature coil of these lines of force let us connect its two ends with our solenoid (Fig. 2,

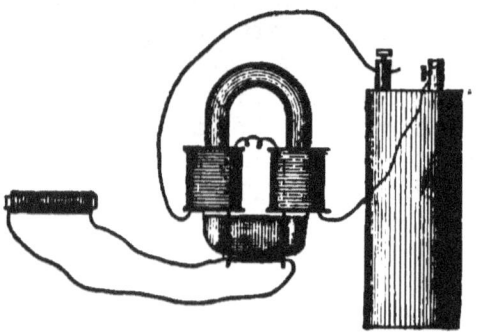

FIG. 19.

page 35). Next take a very small needle that has been previously magnetized and suspend it at its center, so that it will hang horizontally. Use a coarse spider-web for suspension if possible, as even the finest thread is a little stiff and has a twist which will interfere more or less with the action we are looking for. Hold the solenoid up to the needle so that the latter projects for about one-third of its length into the solenoid. If one of the battery wires is disconnected from the battery, make the connection by touching it to its binding post. This, as we have seen, will cause a rush of lines of force through the armature core. Note the behavior of the needle as this rush occurs. It will give a sudden kick, either outwardly or inwardly, showing that the solenoid has momentarily exerted upon it either a force of attraction or one of repulsion. But it is only momentarily, for if the conditions remain the same, the needle, after swaying back and forth a few times, will come to rest again in exactly the same position that it assumed before the lines of force traversed the armature core. The lines of force are still there, but there is neither attraction nor repulsion of the solenoid. Now break the battery circuit so as to suddenly withdraw the lines of force. Another kick will be noticed in the needle, but in the opposite direction. If the solenoid exerted an impulse of *attraction* when the lines of force passed into the armature core, it will exert an impulse of *repulsion* when they are suddenly withdrawn. Next reverse the ends of the armature and repeat the experiment without changing the position of the solenoid or its connections. In the reversed position of the armature the end that was before in contact with the north pole of the magnet will be in contact with its south pole, and that which was in contact with the south pole will be against the north pole, so that the lines of force as they enter and withdraw in the same directions with regard to the magnet have opposite directions with respect to the coil. As they enter, the solenoid will be found to *repel* and to *attract* where they withdraw.

Now modify the experiment a little. While the magnet remains excited jerk off the armature and then replace it. Exactly the same effect will be produced upon the needle by the solenoid as before when the current was alternately broken and closed. Next disconnect the solenoid, and connect in its place in the armature circuit the flat coil (Fig. 6, see page 41), which we employed in our galvanometer. Hold this directly beneath the needle with its wires parallel with the latter. Upon breaking and making the battery circuit or pulling off or putting on the armature we will find that the needle receives momentary impulses which cause it to deviate alternately to the east and to the

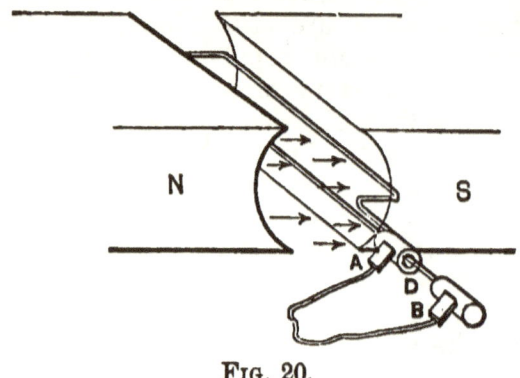

FIG. 20.

west. As with the solenoid, the effect is only momentary and not sufficient in the present instance to cause the needle to swing far, but by timing the makes and breaks, remembering that they give impulses in opposite directions, the needle may be caused to swing through gradually increasing arcs, and finally to rotate completely around on its support.

We will have recognized before this that the action of both the solenoid and flat coil upon the needle was due to an electric current caused by the sudden introduction and withdrawal of the lines of force through the armature coil, and that the direction of the resulting current changed as the number of the lines of force threading the coil is increasing or decreasing. In fact, had we a machine by which the armature could be rapidly approached to or removed from the magnet poles we would have an alternating current generator producing currents in the armature circuit exactly similar to those employed in lighting by alternating currents, and if, further, we had a device by which the reverse currents after they are generated could be changed in direction to correspond with those which both precede and follow them, we would have in all respects a direct current generator. In fact, almost before we have been aware of it, we have actually developed experimentally an electrical generator. It now only remains to develop the details by which the current generated is rendered continuous in one direction instead of alternating, and practically steady instead of pulsatory, as it would be were the currents we have just generated all sent in the same direction. We have also discovered that the electromotive force is not due to the actual number of lines threading the coil, for the same effect was produced when there were no lines—when the magnet was not excited—as when there was the greatest number—that is to say, there was no effect in either case, but the current flowed only when the number of lines was changing. The rule is that the potential difference, which gives rise to the current, is proportional not to the number of lines included by the coil, but to the *rate of change* of the number of lines. In our experiments it was an almost instantaneous change from no lines to the full number we were able to generate, and from that number to none again.

THE CONTINUOUS CURRENT DYNAMO.

In the dynamo of to-day a much more convenient method of varying the number of lines included in the coil is obtained by revolving that coil around an axis in a uniform magnet field.

In Fig. 20 we have a representation of a single turn of wire revolving around one of its sides, as an axis in a uniform magnetic field. In the position shown none of the arrows will be embraced by the coil, but as we revolve it in either direction it gradually will allow more and more lines of force to pass through it, until it arrives at a position at right angles to the one shown, when it will embrace the maximum number, but in this position its motion will be for a moment parallel with the lines of force, so that for a very small portion of its revolution near this position it will cut no more and no fewer lines. Its rate of cutting lines at this point being zero, no electromotive force, and consequently no current, will be generated. As it proceeds around through another right angle the number of lines which the loop will include diminishes, gradually at first, but more and more rapidly, until the position of the loop is directly opposite that shown in the cut, when it again becomes parallel with the lines and for a moment includes no lines, but as it passes this position the loop turns its other side to the north pole, and commences to take in lines from the opposite side—that is to say that with respect to the loop the direction of the lines is reversed. At this point the rate of change of the number of lines embraced by the coil is a maximum, because at one instant there were a few lines going through the coil in one direction, and at the next there was the same number going through in the opposite direction, or there were just that many less than nothing going through in the original direction the second moment. The greatest rate of change in the number of lines embraced by the coil that can possibly occur takes place at this part of the revolution, and therefore at this point in its path the greatest electromotive force is generated. From this point to the vertical position of the loop, at right angles to the lines of force, the rate of change becomes slower and slower, and the electromotive

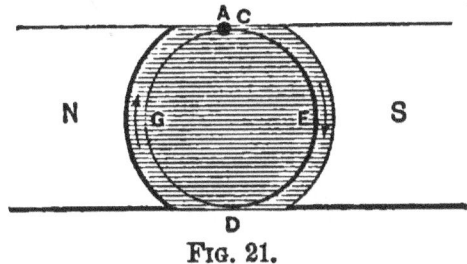

Fig. 21.

force less, until the loop arrives at the latter position, when for a moment there is no change, since its motion is for the time again parallel with the lines of force. No electromotive force is, therefore, developed at this point in the revolution, but from here to the original position, shown in Fig. 20, the rate of change gradually increases again, until it becomes a maximum once more in the position shown. It will be observed, therefore, that, as the coil is revolved between the two poles of the magnet, the electromotive force generated twice reaches a maximum, once when the coil is in the position shown, and again when 180° from this position, and twice becomes zero, viz., when it is in the two positions at right angles to this plane. In each case, as the electromotive force passes through zero, the current resulting changes its direction, so that in each revolution of the coil the current will flow half the time through the outside circuit from A to B, and during the other half in the contrary direction, from B to A.

The more usual way of explaining this generation of electromotive force is to speak of the rate at which a wire represented in section by the dot C, Fig. 21, cuts the lines of force when revolved in a circular path $A\ E\ D\ G$ in a uniform magnetic field represented by the parallel lines. Referring to the figure, when the wire is at C, it is traveling for the moment parallel with the lines of force, and therefore cutting none and generating no electromotive force. As it proceeds around from left to right it cuts these lines more and more rapidly, until at E it is moving at right angles to them, where it cuts them at the maximum rate of its course. At this part of its path the highest electromotive force is generated. From E to D it cuts them less and less rapidly, until it arrives at D, where its motion is again parallel to the lines and no electromotive force is generated. From D to G the rate again increases and from G to C decreases, but the direction of the current during this half of its excursion is in the opposite direction to that resulting from its course in the first half—the changes of direction of the electromotive force or pressure upon which the current and its direction depend taking place as the wire passes through the positions where it cuts no lines, called the neutral positions, A and D.

While the same results are reached by explaining the action of a moving wire in a magnetic field in this way as in the other, it is not strictly a correct explanation, for according to it the generation of electromotive force is made to depend upon the actual cutting of the lines of force by the moving wire. We have seen that this is not a fact, for in our experiment with the electromagnet (Fig. 19) we found that with our coil on B perfectly stationary we generated an electromotive

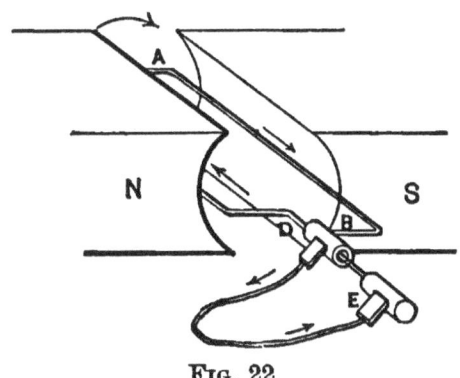

Fig. 22.

force simply by making and breaking the battery connection with the magnet coils, thereby rapidly changing the number of lines that threaded through the armature core, from zero to a maximum, and *vice versa*. In this case there was absolutely no cutting of lines. But, as before stated, the same results follow both methods of explanation, and the latter, although strictly speaking not correct, is simpler, and for that reason will be used hereafter.

Fig. 20 represents an electromagnetic generator in its simplest form.

It is more usual, however, instead of having a

narrow coil revolving around one of its sides as an axis, to employ a larger coil and revolve it around an imaginary axis in its center, as shown in Fig. 22. In this case when the coil revolves in the direction indicated, while the side *A B* cuts the lines in one direction, the other side, *A D*, cuts them in the opposite direction. In the cut the end *B* is represented as terminating in the hollow cylinder *B*, through the axis of which passes the other end of the coil, also terminating in a cylin-

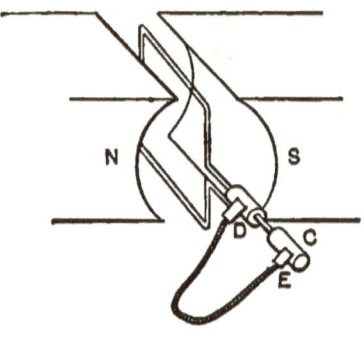

FIG. 23.

der. Upon these two cylinders copper brushes are pressed, to which are attached the two ends of the external circuit *D G E*.

With the coil in the position shown the maximum electromotive force is being generated for the reasons already explained, and the direction of the resulting current will be as indicated by the arrows. When the coil has arrived at the position shown in Fig. 23, it is cutting no lines of force and generating no electromotive force, but immediately after passing this position it commences to cut the lines again, the two sides cutting the lines in opposite directions, however, viz., *A B* has exchanged places with *A D*, the former cutting them from west to east, if in looking toward the north pole we be supposed to be looking north, and the latter now cutting them from east to west, using the same points of the compass. The current will, therefore, be reversed, and will reach its maximum strength when in the position shown in

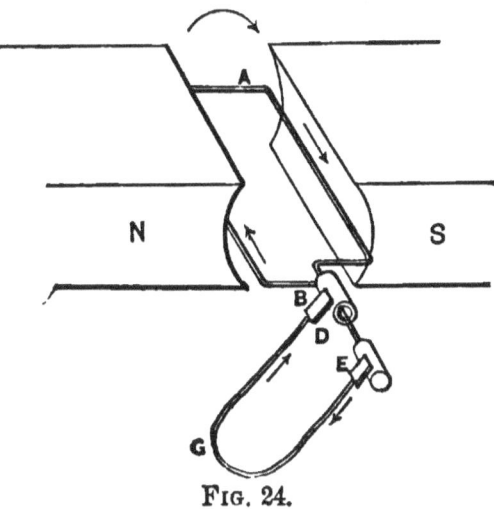

Fig. 24.

Fig. 24. It will again become zero and reverse its direction when the coil has reached a position at right angles to this; and so the changes will follow each other as the coil revolves, repeating the changes with each revolution, becoming zero and reversing its direction each time the coil takes up a position at right angles to the lines of force, and reaching a maximum each time it becomes parallel with them.

We have here what may be termed a typical

machine generating alternating currents, viz., those which are periodically reversing their direction. What is wanted, however, is to generate a current that shall flow always in the same direction in the outer circuit. It is evident that if at the moment the coil arrived in its neutral position (Fig. 23), when for one moment it is generating no current and the next commences to generate one

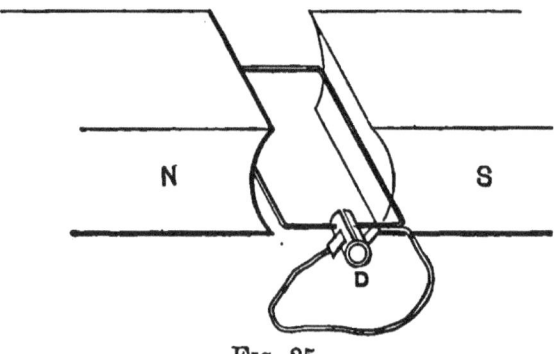

Fig. 25.

in the opposite direction, we should exchange the places of the two brushes D and E, placing E upon the cylinder which terminates the end of the wire B, and D upon the cylinder C, and replace them again in their original positions when the coil arrives at its next neutral position, the current in the external circuit would no longer be reversed, but would continue to flow in the same direction throughout the complete revolution of the coil. A simpler way of accomplishing the same thing, however, is at hand.

Suppose the hollow cylinder D, in Figs. 20, 22, 23, 24 (see pages 78–85), be slit longitudinally into two equal parts, and let one part be connected to each of the two ends of the turn of the wire, as

THE CONTINUOUS CURRENT DYNAMO. 87

shown in perspective in Figs. 25 and 26, and in section to a larger scale in Figs. 27 and 28.

Calling these two halves *m* and *n*, if we place the slits at right angles to the coil, as shown in the figures, and place the brushes *D E* in the positions shown, the direction of the current in the outer circuit will be automatically changed at the proper time. In Fig. 28 the coil is in one of its neutral positions. Just before this *D* has been in contact with *m* and *E* with *n;* the current

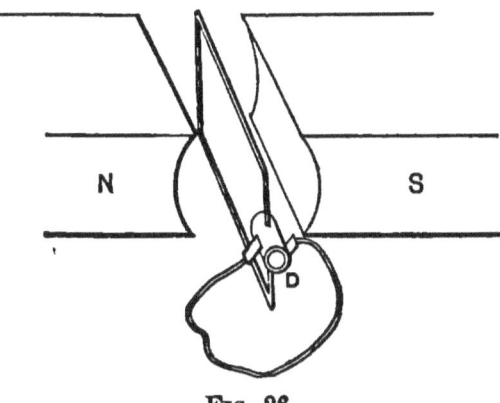

FIG. 26.

at that time was flowing from *n* through *E G D* to *m*. When the coil has arrived at the vertical position shown in Fig. 28, there is no electromotive force generated, but just after it has passed this position section *n* passes from under brush *E* and section *m* passes from under brush *D*. Therefore brush *E* rests upon section *m* and brush *D* upon section *n* (Fig. 29), just as the direction of the electromotive force in the coil changes, so that the current will continue to flow in the same direction in the circuit *E G D* as before.

This arrangement, by which the alternating current is changed to one always flowing in the same direction, is termed a *commutator*, and the line joining the positions of the brushes where the change of direction or commutation takes place is called the " axis of commutation."

We have already accomplished part of our purpose, but not all, for while the current now always flows in the same direction, it is an exceedingly unsteady one, having at one time no electromotive force at all and at another a maximum. We must

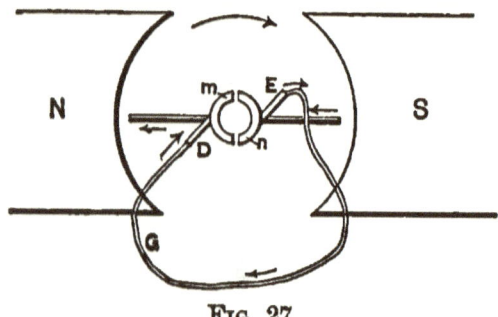

Fig. 27.

have a more uniform current, and to this end let us add another coil at right angles to the first (Fig. 30).

If this second coil is added and the commutator split into four sections, the currents in both coils will be rectified, and as the coil b is in its neutral position while coil c is generating its maximum electromotive force, and *vice versa*, there will be no time when the circuit is without current, for while coil b is contributing nothing, the other coil, c, is doing its best. Referring to the cut it will be seen that each coil will come into action when it comes within 45° of its position of maxi-

THE CONTINUOUS CURRENT DYNAMO. 89

mum effect, and will go out of action when it has passed 45° beyond that point.

While in the position shown the coil c is in a maximum position, and section o is positive and p negative. The current will therefore flow from brush E through the circuit to B. The potential will diminish until the coil has passed through 45°. Then the sections o and p of the commutator and the coil will no longer be in connection with the brushes and the outer circuit, and may be neglected

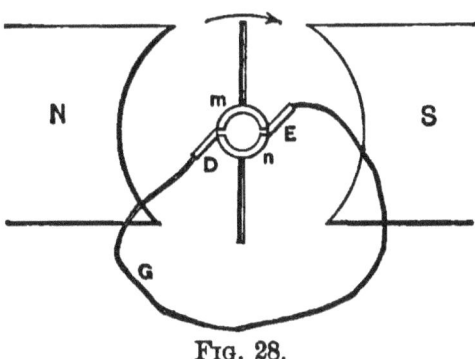

FIG. 28.

for the next quarter of a revolution. When the segments o and p pass from under the brushes, segments m and n immediately succeed them and coil b is connected to circuit. This coil is approaching its position of maximum effect, and therefore its potential difference is increasing, and will continue to increase as it passes through an angle of 45°, when it reaches its maximum, and will then decrease as did coil c for 45° more, until like the other coil it is cut out of circuit by its commutator segments passing from beneath the brushes. Thus a decreasing electromotive force of a strength due to a position of one coil 45° beyond its position of maximum effect is succeeded by an increasing electromotive force from another coil due to its position of 45° in front of its position of maximum effect, and we have a current still varying in strength, but only between that which would result from a position of the coils of maximum effect and that which would result from a position 45° removed from it instead of one varying between a maximum and zero.

If we double the number of coils again and divide each of our commutator segments into two, the fluctuations will become still less, and so, by multiplication of coils each terminating at both ends in commutator segments which come under the brushes and leave them at a less angle from the position of maximum effect, the resulting current will vary between narrower and narrower limits and gradually approach uniformity of strength.

INCREASE OF ELECTROMOTIVE FORCE.

We observed from our experiments with the solenoids that two turns affected the needle more than one turn did, and it has been stated that the magnetizing effect of a coil or the magnetomotive force of an electromagnet was proportional to the product of the current flowing, into the number of turns in the coil, or the ampere turns, as we have called this product. The converse of this is also true. If a coil of one turn of wire such as has been discussed heretofore revolving in a given magnetic field at a certain speed generates an electromotive force of say one volt, a coil of two turns or a coil of three turns revolving at the same speed in the same field will generate an electro-

INCREASE OF ELECTROMOTIVE FORCE. 91

motive force of double or treble as much. That is to say that since a given current passing two and three times around a bar of iron or space will generate two and three times as many lines of force as it will if it passes around but once, so if a coil of one turn cutting the lines of force at a given rate generates a certain electromotive force, the same wire bent into a coil of two and three turns will under like conditions generate two and

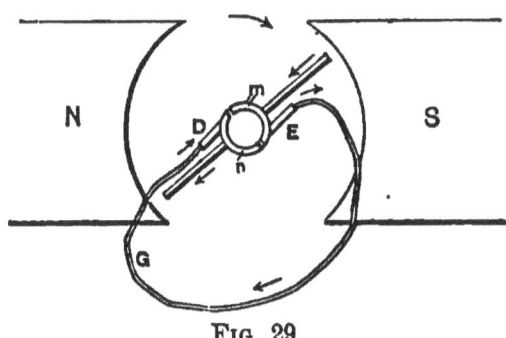

Fig. 29.

three times as many volts. Thus if a coil such as is represented in Fig. 31 (see page 93) be substituted for the coils of a single turn represented in the previous illustrations, it will at the same speed of revolution generate twice as much electromotive force, because in doubling the wire it is equivalent to doubling the rate at which that wire cuts the lines of force—which alone determines the electromotive force that will result. This being the law, many other means of increasing the electromotive force at once suggest themselves. If we revolve our coil faster, its rate of cutting will be greater, or if we increase the number of the lines, the rate of cutting will still be faster, even though the speed of revolution be not increased. This latter statement suggests two other means of increasing the electromotive force, for we can increase the number of lines of force in our field in two ways —either by increasing the magnetomotive force of our magnet by increasing the number of ampere turns, or by decreasing the magnetic resistance of the air space in which our coils revolve. Since

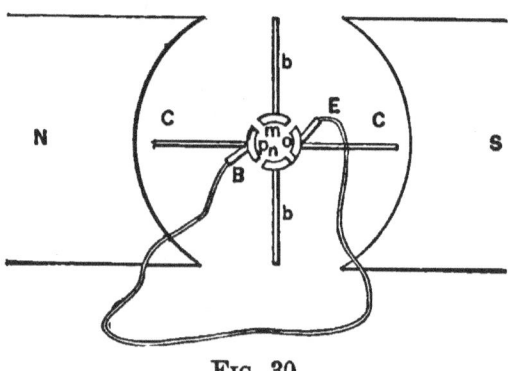

FIG. 30.

our coils are of a certain size, we cannot bring the poles of the magnet closer together, as we did in Fig. 12 (see page 62), and still leave room for our coils to revolve, but we can wind our coils on a cylinder of iron, which is the best conductor of the magnetic lines that is known, and thus by greatly reducing the resistance in this gap greatly increase the number of lines that our coils will cut at a given speed, and thus increase enormously our electromotive force. And this method is always employed in dynamo construction, because it results in an enormous saving in copper, since with one turn of the wire on an iron core a greater number of volts will be generated than with very

EDDY CURRENTS IN ARMATURE. 93

many turns of wire without the iron revolved at the same speed. Nor is any expense spared to have this core of the softest and best iron for the purpose, for the saving in other directions where the best iron is employed in the armature core more than counterbalances the additional cost of the extra quality.

EDDY CURRENTS IN ARMATURE.

Those who have seen generator or motor armatures in process of construction will have noticed also that they are not made of one solid piece, but are built up of a great number of disks of thin

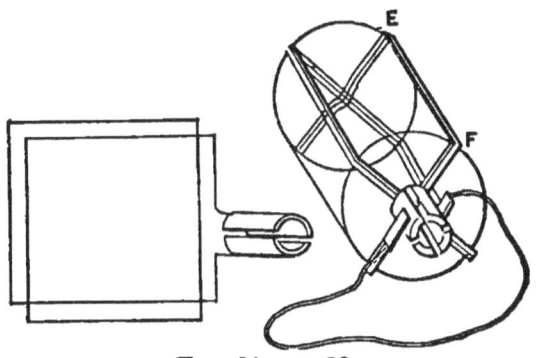

FIG. 31 AND 32.

sheet iron, each of these disks being insulated from its neighbors usually by thin sheets of paper.

This lamination, as it is called, is not for the purpose of increasing the capacity of the iron to carry lines of force, but is for the purpose of permitting the core to rapidly change the direction of its magnetization. As will be seen later on the armature becomes a magnet whose poles always have a constant position with regard to the field magnet poles—that is, they are stationary with regard to the latter—but since the armature is revolving, they are constantly changing with regard to a fixed point on the armature. This lamination greatly facilitates the rapid shifting of these poles, which if retarded, as would be the case even with the best iron if it were solid, would give rise to eddy, or Foucault currents, as they are called, in the iron itself, which would be at the expense

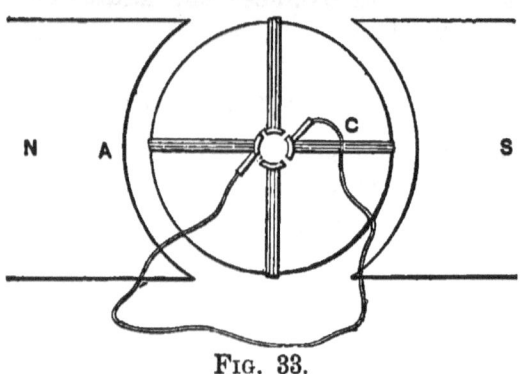

FIG. 33.

of the engine which drives the armature, would contribute nothing in return to the output of the dynamo, would cause the armature to heat up to an abnormal degree so as to perhaps destroy the insulation on the armature wires, and cause irregularities of working too numerous to mention at this point.

Figs. 32 and 33 represent an armature consisting of two coils of two turns wound on an iron core at right angles to each other. It will be seen that in Fig. 33 when the armature is inserted between the pole pieces $N\ S$ the distance which the lines of force have to pass through air is reduced to the two narrow spaces which need only be wide enough to permit the armature to revolve rapidly without allowing the coils to strike against the faces of the field magnets. Of course as the armature revolves the wires on its surface tend to fly outward, and sufficient room must be allowed to provide for this. Other precautions are taken to prevent these wires, $E\,F$, Fig. 32, from flying outward or from moving in any way from their position. In small motors and dynamos the most common way to prevent this is to bind them down tightly to the core with a number of windings of wire.

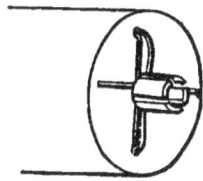

FIG. 34.

In larger machines it is now becoming very common practice, instead of winding the wires on the outside of the cylindrical surface of the core, to carry them through slots cut in the iron itself, each coil, whatever the number of turns, occupying a slot by itself.

Fig. 34 represents an armature thus wound. This method where the size of the machine permits has several advantages, some of which are mechanical, while others are purely electrical. One of the chief advantages is that it is impossible for the wires to move, and another is that the air space between the armature and the pole pieces of the field magnet may be made considerably smaller, for it is evident that the space taken up by the windings of the wire on the outside of the armature offers quite as high resistance to the magnetic flow as does that additional space which must necessarily be left for clearance.

CHAPTER X.

SHIFTING OF THE ARMATURE WIRES.

WE have laid some stress upon the injunction that the windings of the armature must not be allowed to move. It is evident that if they are free to move ever so little it will be almost sure to result in the breaking of the insulation with which they are protected, and this being destroyed, even over spaces the size of a pin head, the current will pass from wire to wire of the coils, or from wire to armature core, instead of passing out through the commutator and outer circuit, and this new path being short and of low resistance, the flow will be large, or if not large at first rapidly become larger, and the armature burns out. In fact, this slight movement of the coils is one of the most fruitful causes of burning out of armatures.

There are three agencies constantly at work when either the dynamo or motor is in operation which tend to cause movement of the coils and hence their ultimate destruction. One has already been referred to—the tendency of the wires to fly off from the core when the latter is revolving rapidly, which is counteracted by binding them down tightly to the core by bands or coils of fine wire. Another is that when a dynamo is doing work, and the coils are passing rapidly through the lines of force, the latter act like material obstructions to their motion with the core and tend to push them off sideways. It is very much as though we were revolving the armature rapidly in a tank of water; considerable friction would result between the water and the wires on the surface of the cylinder, which would tend to strip the former from the latter, and this tendency would increase with the speed. One can form an estimate of how great this stripping force is when he realizes that practically all of the force exerted by the engine in driving the armature is required to overcome it. That is to say, that when an engine is exerting say 100 H. P. in driving an armature there is practically 100 H. P. being exerted upon the windings of that armature tending to push them off sideways or to strip them from their position. This stripping force is divided among the coils, and increases for each coil with the number of turns in the coil. In the case of the dynamo this friction acts as a drag upon the coils, tending to prevent them from passing from under the pole pieces of the magnets. In the case of a motor the action is reversed, and it manifests itself as a pull upon the wires as they approach the pole pieces.

It is bad enough when this tendency to strip is always in one direction, for, as the force exerted is constantly varying with the load on the machine, there is a tendency for the coils to move accordingly, which if allowed to occur must inevitably result in friction between the wires themselves, or between the wires and the core, which will wear away the insulation in places and cause a short circuit and a burn out, but it is still worse where the strain comes first in one direction and then in the opposite, as is the case with motors which are constantly reversed, as are street car motors. One of the trials to which street car motors are peculiarly subject is the tugging in one direction at the wires on the armature on starting up, and the tugging at them again in the opposite direction whenever the car is reversed, and it is much aggravated if these operations are performed too suddenly. This is one of the reasons why motormen are always instructed not to turn on the current too rapidly in either direction, and why provision is made in the controlling switch so that the current cannot be turned on or off or reversed at full strength. The deterioration of reversing motors (street car motors) due to this cause, even with the best of care on the part of the motorman, is slow but *sure*.

There are several methods of obviating this tendency of the wires to shift, with which, however, the motorman has nothing to do. One is that which is resorted to to prevent the wires from flying off tangentially, previously described, of tightly binding the coils to the armature by metal bands or wrappings of wire, and the other is by imbedding the coils in channels or slots in the armature core as shown in Fig. 34 (see page 95). This latter method, however, is seldom practicable in small motors such as are used on street cars, because of lack of room for the required number of coils, but in large generators and motors of many hundred horse power it is becoming a favorite method, not only on account of its efficiency for the purpose, but because this method of building the armature has other advantages to recommend it.

In street car motors, however, where of all cases there should be the best possible provision against these strains, we have to rely upon the least efficient method, viz., of holding the wires in place by the binding wires referred to. The motorman who has the best interests of his employers at heart will therefore refrain from jerking his car, for he will remember that these jerks all come upon the armature wires and will hasten the destruction of his machine.

The third agency which tends to cause the wires to move from their position upon the armature is heat. As everyone knows, metal expands with increase of temperature. Now the coils of an armature are wound as tightly as possible, but if the temperature of the wire be raised far above that at which it was wound the wire will become appreciably longer, and the coils, which were tight when cool, become loose when hot. Upon cooling again they contract, but this very expansion and contraction has caused a motion of the various convolutions of the wire relative to each other and to the core which may be more or less harmful by causing abrasion; but the heating is still more harmful for the reason that it aggravates the other tendencies toward movement of the coils already mentioned, for if these strains are brought to bear upon a coil that is already loose it will be readily seen how much more harmful they may become. The only remedy for this is to avoid overworking your machine. Over this remedy the motorman has almost complete control. Starting up or reversing too suddenly, or taking a heavy load too rapidly up a grade, are all examples of practices that should be prohibited, because in all of them

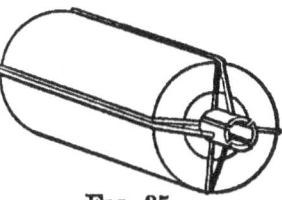

FIG. 35.

the wires are overworked and likely to heat even to burning out. Running at a high speed on a level track is *not* overworking the motor, however, and may be indulged in with impunity, almost, so far as the heating of the motor is concerned, so that it is much better to lose time on grades and make it up on the level stretches than to save time at the expense of the motor where the hardest work is being done, viz., in starting and in ascending grades and going round curves.

OPEN AND CLOSED COIL ARMATURES.

In all of the preceding cuts where the armature is represented as having two coils and four commutator segments it will be observed that, after the coil has passed 45° from its position of maximum activity, its commutator segments pass from under the brushes, and the outside circuit remains disconnected from the coil until the latter has revolved through 90° and again approaches within 45° of its next position of maximum activity. This occurs twice during each revolution of every coil. That is to say, twice during every revolution of the coil it is open and contributes nothing to the outside circuit. An armature wound in this way is called an open coil armature. Such construction is now never found upon street railway apparatus of any kind, either motors or generators, but is thought to have some advantages for arc lighting dynamos, and is the method employed on both the Brush and Thomson-Houston arc dynamos.

If, instead of having connected the coil so that it was out of circuit except when generating a certain potential, we had connected it so that it

OPEN AND CLOSED COIL ARMATURES. 101

would be always in circuit except when generating no potential at all (the moment of reversal, when it would be short-circuited by the brush), we would have taken advantage of the small electromotive forces which we threw away in the other arrangement by cutting out the coils before they had reached and after they had passed 45° of their positions of maximum activity.

Fig. 35 shows an armature wound in this way. By comparing it with Figs. 32 and 33 (see pages 93-4), it will be seen that the only difference between the two is that the two coils are really the one a continuation of the other, but each has separate connections with its own commutator segment. Each coil is therefore always connected with the outside circuit—directly through its commutator segments when the latter are under the brushes, and indirectly through the other coil and its commutator blocks when its own have passed from beneath the brushes. Armatures thus wound are called *closed* coil armatures, and are the kind now universally employed both for generators and motors in street railway equipments and for generators in incandescent lighting stations.

CHAPTER XI.

DRUM AND RING ARMATURES.

BESIDES the two classes of armatures, open and closed coil, above outlined, there are also two other classes of importance, each of which may be either open or closed coil. Reference is here made to what are termed "drum" and "ring" armatures. Those heretofore have represented the wire wound lengthwise over a cylindrical or drum-shaped core. They are therefore for this reason termed "drum" armatures, and, as before stated, may be either open or closed coils.

Instead of winding our wire on a cylinder we may wind it on an iron ring. Fig. 36 represents an elementary open coil armature of this kind corresponding to the open coil drum armature represented in Fig. 20 (see page 78). Another form, with two coils, is shown in Fig. 37, which corresponds very closely with the drum winding in Fig. 31 (see page 93). Fig. 38 is the counterpart in the ring type of armature of Fig. 32 (see page 93), in the drum winding.

Fig. 39 (see page 107) shows a four-part ring armature of the closed coil type. In this can be shown more clearly than was possible in the drum armature drawing (Fig. 35, see page 99), how the various coils are connected together and to the commutator in the closed coil type of armature.

Of course in order to have the greatest output

possible, whether the armature be drum or ring winding, open or closed coil type, the whole of the surface of the core is overwound with wire, and this winding is divided up into as many coils as the designer desires, the two ends of each coil being connected to appropriate commutator segments, and if of the closed coil type also to the ends of the adjacent coils. In the closed coil ring arma-

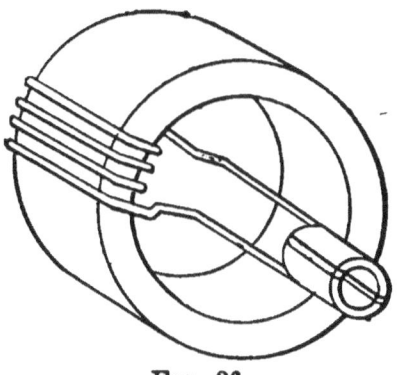

FIG. 36.

ture, Fig. 40 (see page 107), shows how the winding is really one continuous coil all the way around the ring, its two ends finally being joined together. Then the number of turns of wire that shall constitute a coil having been determined on, the continuous coil is tapped at those intervals and connected by means of a short piece of insulated copper wire with a commutator bar or segment, as is also shown in Fig. 40.

Now since the core of a ring armature may be regarded as a cylinder, just like that used in the drum armature, with its center cut out, it is evident that the former must be given a larger diameter in order to give it the same capacity for carrying lines of force, and the latter, instead of cutting directly across from pole to pole of the magnet as they do in the drum armature, in the ring armature follow the iron in preference to the shorter path through the air space in the center (Fig. 41).

The fact that the ring armature must be of larger diameter than the drum armature for the same capacity prohibits its use where the greatest power is required in the smallest possible space, and since in the street car motor this is a first requisite, we seldom see the ring armature employed. On the other hand, in machines of great size, such as street railway generators, it is more frequently employed than the drum armature, chiefly for structural reasons.

CONSEQUENT POLES AND MULTIPOLAR FIELD MAGNETS.

Heretofore in considering magnets we have dealt only with those of a straight bar or horseshoe shape, having two poles, or the closed iron ring, with the closed magnetic circuit, having *no* poles at all. Before proceeding further it may be well to speak of two other forms frequently used both in dynamos and motors. Suppose we have a closed magnetic circuit, as in Fig. 42, and place upon it on one side a coil with a current passing around it so as to make its upper portion a north pole and its lower portion a south pole. If nothing more were done, the lines of force would flow around—part of them through the armature, and perhaps a greater portion would flow around through the other leg on the right. All these latter would be lost, so far as our armature is concerned. But if we place a second coil on the leg

CONSEQUENT POLES.   105

so as to make the upper part of the magnet a north pole also, it will drive these lines back and send additional lines of its own in the same direction. In fact, the effect of the two coils working in opposition is to make two north poles adjoining each other above and two south poles adjoining each other below, as shown in Fig. 42. As like poles repel each other their lines cannot pass each other, but are forced to take the path indicated.

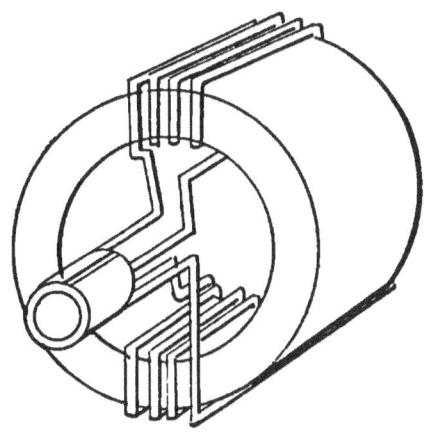

Fig. 37.

When the direction of the current in two legs of a magnet is such as to bring two poles of the same kind together, the latter are called *consequent* poles.

It is possible, therefore, in any piece of soft iron, such as a closed ring, for instance, by dividing the winding up into any number of coils, and passing the current through adjacent coils in opposite directions, to make a magnet of as many poles as is desired. In a closed ring, or equivalent shape, with a single coil, if there be no leakage of lines across from one side to the other, there will be no poles at all. If there be two coils upon this ring through which the current passes, so as to compel the lines of force to pass around in opposite directions, since the lines induced by one coil cannot pass through the other coil, the lines due to both coils will have to seek another path, and will emerge from the ring somewhere between the two coils, and jumping across to the other side along the path offering the least resistance, re-enter

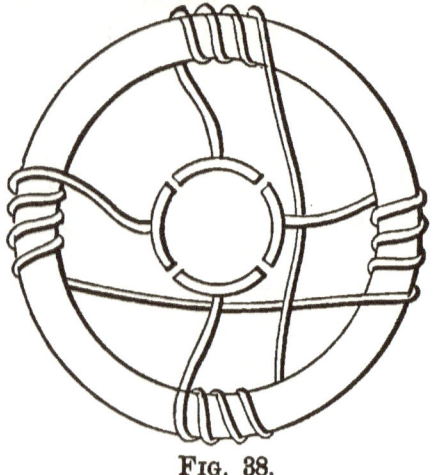

FIG. 88.

the ring again, producing consequent poles, as already described and illustrated in Fig. 42.

If we take an iron ring and wind it, as in Fig. 43, with four coils, $A \ B \ C \ D$, connecting them up so that the current will travel around the ring in the directions indicated by the arrows, we find that the directions of the currents in coils B and C, if looked at from a point in the ring midway between the two, will be in the opposite direction to the travel of the hands of a clock. Therefore both currents tend to make the space between them a north pole. Moving along now and placing ourselves in the ring again between coils *A* and *B*,

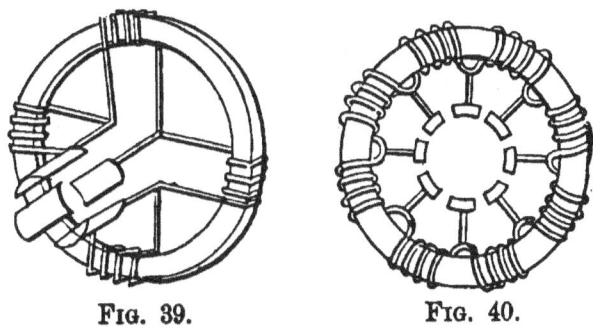

Fig. 39.   Fig. 40.

and looking toward *B*, the current is found to be flowing around the ring in the same direction as the hands of a clock. That end of the coil *B* at which we are looking is therefore a south pole.

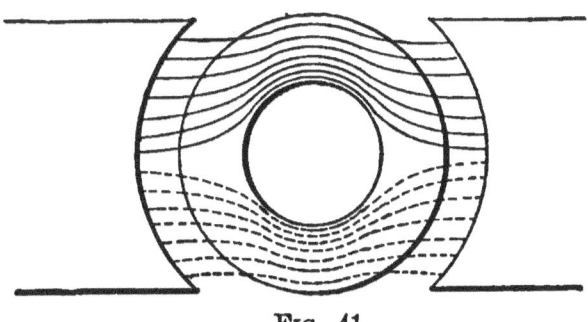

Fig. 41.

Turning around and looking at the end of coil *A* nearest to us, we find that the current as we see it from this point of view is also circling around in the direction of the hands of a clock, so that this end of *B* must also be a south pole. We might have known this without looking at it, for if the other end of *B* was a north pole this end must be a south pole. We find that the two adjacent ends of the coils *A* and *B* are south poles, so the space between them will be consequent south poles. In like manner we will find consequent north poles between *B* and *C* and consequent south poles

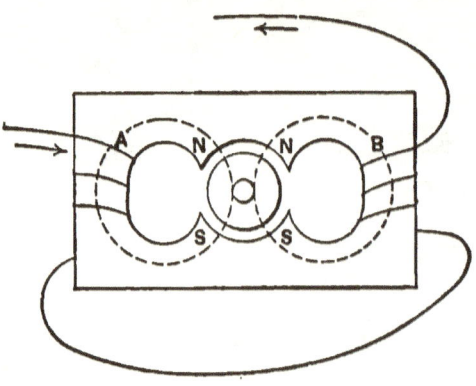

FIG. 42.

between *C* and *D* as lettered in the diagram. Thus we have a magnet with four poles—two south poles and two north poles. A magnet having but two poles, such as described in the earlier chapters and also in Fig. 42 (for the two consequent poles are counted as one), is called a "bipolar" magnet, and a magnet having more than two poles, as shown in Fig. 43, is called a "multipolar" magnet.

CHAPTER XII.

MULTIPLE ARC AND SERIES ARRANGEMENT.

As observed in Fig. 43 the current, before it comes to the magnet coils, divides—part of it going through coils B and C, and part of it going through A and D, and then the two parts unite again after having passed through their respective coils. It is evident that instead of dividing the current into two circuits, each half going through two coils in succession, we might have divided it into four circuits, and have placed each coil in a separate circuit, or we need not have divided the current at all, but have compelled the whole of our current to pass through first A, then D, then C and finally through B. Fig. 44 represents the latter arrangement. When current is supplied to several coils or lamps or other electrical devices or groups of the same, so that each device or group receives a certain fraction of the whole current independently of all the others, viz., so that the current which passes through one device or group does not pass through any of the others, as would be the case were we to divide our circuit into four in one of which each of the four coils are placed, these four coils would be said to be in "*multiple arc*," or "in multiple" with each other, or "in parallel." In Fig. 43 the current divides between two circuits, the current of one of these circuits passing through the group of coils A and D, and the current of the other passing through the group *B* and *C*.' These two subordinate circuits are therefore in multiple arc with each other, and the groups of coils supplied by each circuit are said to be " in multiple arc " or "in parallel " with each other, because none of the current which passes through the coils *A* and *D* subsequently passes through the coils *B* and *C*.

If, however, the current reaches its translating devices (the term " translating device " is used to

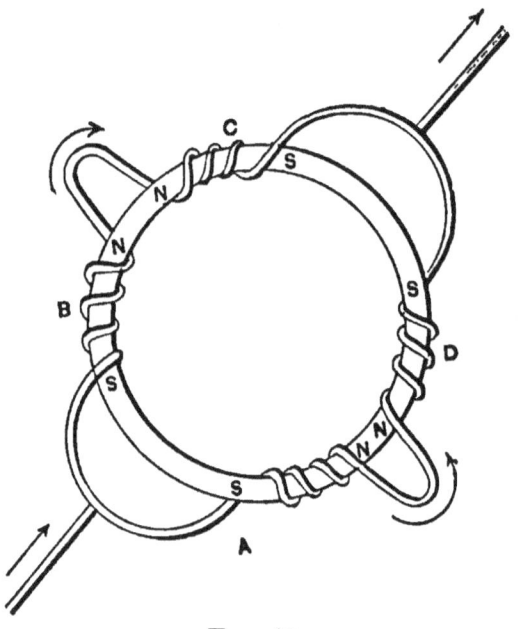

Fig. 43.

designate anything through which the current passes or which is operated by the current either directly or indirectly) not at the same time or independently as in previous example, but passes through them in succession, so that the same current which passes through the first one also passes through the second and the third and so on until all are thus supplied, the various devices are said to be "in series" with one another. Thus in Fig. 43, while coils A and D, considered as a group, are on a separate circuit from B and C, and the current which goes through A and D does not afterward pass through B and C, and the two groups are in multiple arc with each, still, if we regard the coils separately, the same current which passes through A subsequently passes through D, and that which first passes through B also passes through C.

A and D are therefore in *series* with each other, and B and C are likewise in series with each other, but if we arrange our coils as in Fig. 44, so that the same current which passes first through A and then through D also passes through C and B in the order named, the four coils are then said to be in series with one another. If the circuit divides before it comes to A, one branch going around the four coils, as just described, and the other branch supplying in similar manner the coils of another field magnet, then the two groups of field magnets will be in parallel with each other, while the various coils in each group are in series with the others of the same group. This is exactly a similar case to that represented in Fig. 43, where the group of coils A and D, while its constituents, A and D, are in series with each other, is in multiple arc or in parallel with the other group of coils, B and C, whose constituents, B and C, are in series with each other.

Such an arrangement as this, where the translating devices are divided up into groups, the members of each group forming a series, but the various groups being in multiple arc with each other, is called a "multiple series" arrangement.

If in Fig. 44 the current, after having passed through all four coils, is carried to another group of magnets or coils, these two larger groups will be in series with one another.

Figs. 45, 46 and 47 represent respectively the series, the multiple arc and multiple series arrangement for translating devices. There is also

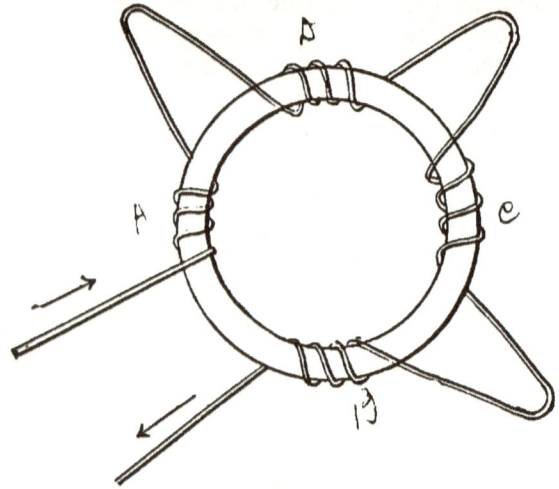

FIG 44.

a fourth arrangement, represented in Fig. 48, which is sometimes called the "series multiple" arrangement.

It is well to fix thoroughly in mind the characteristics of the series, multiple arc and multiple series arrangements, as they will be frequently referred to hereafter, and a thorough understanding of them is essential to an intelligent reading of what follows. To facilitate this understanding it may assist us if we remember that the incandescent lamps in a street car are arranged five in series. The street cars themselves are in multiple arc with each other, as are also the generators at the power house if there be more than one connected to the same feeder. Most of the incandescent lamps except those in street cars are arranged in multiple arc, whereas almost all the arc lamps used for street lighting are arranged in series.

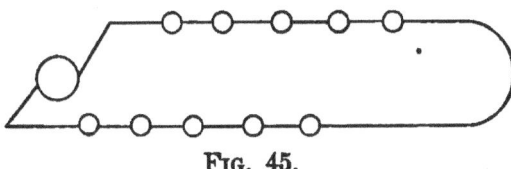

FIG. 45.

The motors under the cars are connected up in various ways according to the position of the controlling switch and the system of equipment employed; sometimes the two motors are in multiple arc with each other, and sometimes they are in series, and the coils of their field magnets are thrown into various combinations of series, multiple series and multiple arc, each arrangement being employed for some specific purpose.

CURRENT CHARACTERISTICS OF MULTIPLE AND SERIES ARRANGEMENTS IN GENERATORS.

If two pumps standing side by side pump water into the same main, they will not increase the pressure of the water, but will increase the amount of water only. So if two dynamos are connected in parallel or multiple arc with the same trolley or feeder wire, the potential of the resulting current will not be increased, but the number of amperes which may be drawn from that circuit, or the number of cars or lights or other translating devices that may be supplied, will be equal to the sum of those that could be supplied by both separately.

If, however, one of two pumps delivers to the other so many gallons of water per minute at a pressure of say 100 pounds per square inch, and

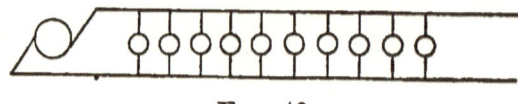

Fig. 46.

the second pump receives that water into its cylinders at that pressure and passes it on to the main with an additional pressure of say 50 pounds per square inch, the amount of water pumped into the main will not be increased by reason of the second pump, but its pressure will have been

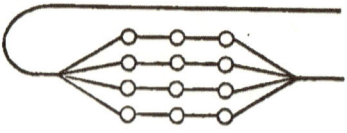

Fig. 47.

increased so as to equal the sum of the pressures imparted by both pumps, in this case equal to 150 pounds per square inch.

So if one dynamo pumps current into another which passes it on to the line wire, or, in other words, if two dynamos are connected in series, the number of amperes that can be drawn from that circuit will not be greater than if but one machine were working, but it will be at a pressure equal to the sum of the pressures impressed upon the current by both.

If in our power house we have a 500-volt dynamo capable of furnishing sufficient current to operate ten cars, and it becomes necessary to double our equipment of cars, we would put in another 500-volt dynamo of the same capacity and connect it up in multiple with the first one, and the question is solved.

If, on the other hand, we had a 100-volt incandescent lighting dynamo, and wished to supply a 500-volt circuit for street car purposes, we would have to connect up others in series with it whose potentials were such that when added together and to that of the machine already installed they

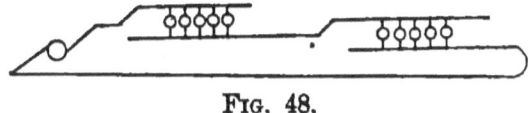

Fig. 48.

would equal 500. Thus we might add 4 more machines of 100 volts each, making 5 machines in series whose combined electromotive forces would be 500, or 2 machines of 200 volts each, or 1 machine of 100 volts and another of 300 volts. But if the original machine only had a capacity of say 100 amperes, the 5 connected in series would have no greater output.

The rule, therefore, may be laid down that a combination of generators in multiple arc gives an output in amperes equal to the combined output of the several machines, but with no increase in pressure, while a combination of generators in series gives a pressure equal to the combined pressures of the several machines, but with no increase in amperes.

CHAPTER XIII.

CURRENT CHARACTERISTICS IN TRANSLATING DEVICES.

When there are no cars on the line, or the trolley wire is not otherwise connected with the rails or the ground, the resistance between the two is infinitely great, and no current will pass from one to the other. If one car be now put into operation, one path will be opened to the current from the trolley wire to the rails. If now a second car and a third be placed in operation, a second and a third path will be opened, and if the resistances of all three of these paths be the same the resistance between the trolley wire and the rail when two cars are in operation will only be one-half, and when three cars are going one-third what it was when but one car was receiving current; the pressure will remain the same with the addition of cars in multiple; but as each car put into service opens a new path for the current between the trolley wire and the rail, the number of amperes of current that will flow will increase with the number of cars—will be twice as much for two and three times as much for three cars as for one.

A parallel case is found in the consumption of water. Supposing we have a line of water mains connected with a reservoir 500 feet high. The pressure of water in the mains at their lowest point would be that due to 500 feet of head. Now suppose we have a number of small faucets tapped into the main at its lowest point. When they are all turned off, the strength of the main, or its resistance, while not infinitely great, as in the illustration of the trolley wire, is still sufficiently great so that no water can flow out at any point. Open one faucet and a single path of, we will say, unknown resistance is opened, and a stream of so many quarts or gallons per minute will flow out onto the ground. Open a second faucet and another path of equal resistance will be opened, and double the quantity of water will flow. The same quantity would flow if, instead of opening a second faucet of the same resistance as the other, the first one were closed and another one of double the capacity or offering half the resistance were turned on in its stead. Therefore, while by opening one small faucet we have decreased the resistance of the pipe from something more than sufficient to keep back water under 500 feet pressure to some definite quantity which is less than that, by opening two of the same size we have divided the resistance remaining by two, and by turning on a third it will be reduced to one-third what it was when but one was turned on. But no matter how many faucets we open, nor how much water runs out (provided, of course, it does not exceed the capacity of the main or the reservoir is not emptied), our reservoir remains 500 feet above us, and the pressure will remain the same at the faucets.

We may make the resemblance between the electric flow and the flow of water still more striking by supposing that each faucet connects with a little water motor which, when working to its fullest capacity, requires a stream of water as large as that which can run out of the faucet when turned on full under 500 feet head.

By turning on the faucet part way, with 500 feet head of water behind it, enough water will flow to enable the motor to do a fraction of its maximum work, turn it further it will do more, and finally turn it on full and it will do the most that it is capable of doing. A second and a third water motor may be attached to other similar faucets and be caused to do varying amounts of work by turning on more and more water, but whether the motors are doing little or much work, using little or much water, the pressure in the main will not decrease until so many motors are attached as to require for their operation more water than the main can carry. If the capacity of the main be exceeded, it will be necessary to lay a larger main, and if this be fed by two pipes of the same size—the original pipe from the same reservoir and another one from another reservoir at the same height—the pressure on the main will not be increased, but its capacity for running water motors will be doubled.

While there will have been no water consumed by the water motors, for the reason that just as many gallons will flow away from the motor as was delivered to it, that which flows away has lost its pressure. If, for example, it required just 10 gallons per minute under 450 feet of pressure to drive each motor at its fullest capacity, and the motor received that amount of water at 500 feet pressure, it may be said to absorb the pressure due to 450 feet, and the water will flow away from the wheel at the rate of 10 gallons per minute, but having a head of but 50 feet. That is to say, a pressure equivalent to that exerted by a column of water 450 feet high will have disappeared in operating the water wheel. A second wheel requiring a pressure of 450 feet head placed below the first wheel could not be operated, because the remaining head would not be sufficient, and if the two were placed together in this manner—in series—neither of them could be operated to their full capacity, because when thus placed their combined resistances would be opposed to the water—that is to say, the resistance of both would be double that of one, and the amount of water that would pass through either would be only half as much as when there was but one. If we increase our pressure, however, more water will flow through our faucet and from one motor to the other, until when we have doubled our pressure, or connected our main to a reservoir 900 feet high, both motors will be operated at their maximum capacity. The water running away from our last motor will then have no pressure, each motor having absorbed a head of 450 feet; the same amount of water, 10 gallons per minute, will, however, be found to have passed through, but it is no longer capable of doing any work.

Thus we see that by operating translating devices in series we *consume* pressure (volts), and not current (amperes), which is just the reverse of what takes place in the generating station where we connect generators in series, thereby increasing the volts with the number of machines, but gaining nothing in the way of amperes.

Take the case of an incandescent dynamo giving current at about 100 volts. Each 16 C. P. lamp requires when working at its normal rate about $\frac{1}{2}$ ampere of current at 100 volts. That is to say, its resistance is such that it requires a pressure of about 100 volts to force sufficient current through the filament to heat the latter to a white heat. If our dynamo had a maximum capacity of 100 amperes at 100 volts, and our lamps required $\frac{1}{2}$ ampere each, 200 lamps could be fully supplied by the circuit if placed in multiple arc with one another, because there would be enough current to go round and it would be delivered to each lamp under the same pressure. Its capacity, however, would then be exhausted. If we connect up a second dynamo of equal capacity to the same circuit in multiple arc with the other, 200 additional lamps could be supplied in the same way. But supposing we place 2 lamps in series across the circuit: their resistances would be added, and at 100 volts pressure but half as much current could flow, viz., $\frac{1}{4}$ ampere. The lamps, we have stated, required $\frac{1}{2}$ ampere each, hence if 2 were placed in series neither would be heated sufficiently to give much light. If, however, we connect our dynamos in series so as to double the electromotive force or pressure of the circuit, the required quantity, $\frac{1}{2}$ ampere, would be forced through the combined resistance of the 2 lamps, and both would burn at their normal candle power. A single lamp, unless of double the resistance of those heretofore used, could not be used on this 200-volt circuit, as the pressure would be so great as to immediately break it. For the same reason we cannot put 100-volt lamps in our cars in multiple arc arrangement, for they would all be destroyed by the 500-volt pressure, as fast as they could be placed in their sockets. But if each lamp absorbs 100 volt of pressure, and each added to a series consumes the same number of volts, if we place 5 lamps

MULTIPOLAR FIELDS.

in series on the usual 500-volt circuit, there will be just enough pressure to give each of them what is required to operate it at normal candle power. We are thus limited on street cars operating on the usual 500-volt circuits to 5 lamps placed in series. If we want more, we must arrange another series of 5, for if we placed 6 or 7 on the first series each would only receive $\frac{1}{6}$ or $\frac{1}{7}$ of 500 volts, which would not be sufficient to illuminate them to full candle power, and if we placed but 1 or 2 on the second circuit, in the case of the single lamp it would receive the full pressure of 500 volts, and in the case of 2 lamps, each would receive 250 volts, which in both cases would be entirely too much. We might, of course, make up the second series of 1 or 2 lamps and dead resistances equivalent to the resistance that would be offered by the additional lamps required to make up the series of 5, but these dead resistances would consume as much energy as the lamps they replaced and give no useful return, so that it would be much better to complete the second series with lamps than with dead resistances. If we had a second series of 5 lamps, this second series would be in multiple with the first.

The same reason which compels us when using 100-volt lamps on a 500-volt circuit to place 5 of them in series would compel us in case our circuit was at 1000 volts to use 10 lamps in series.

MULTIPOLAR FIELDS.

Referring to Fig. 43 we see how an unbroken iron ring may be wound so as to have four distinct poles. In practice it is desirable to have as little air space between these polar surfaces and the armature as possible, and for this purpose the ring which is to become our multipolar magnet is cast with extensions, called pole pieces, extending inwardly, and the magnetizing coils are wound upon these extensions. Fig. 49 (see page 124) shows this arrangement for a four-pole field, also the directions taken by the lines of force. It is clear that in a four-pole magnet we have the exact equivalent of two simple magnets, and a wire on the surface of an armature revolving in this field will pass four poles in one revolution. As it sweeps by the first north pole it will generate a maximum electromotive force in a given direction. This will decrease until it gets halfway between the north and south poles, where it will become zero and change its direction. That is, it will come to the point where the currents must be commutated in order to maintain them in the same direction. From this point until it passes the adjacent south pole the generated electromotive force will be increasing. It then decreases until it is zero at a position halfway beween the south pole and the next north pole, where it must be commutated again, and so on until the armature has made a complete revolution. Thus in a four-pole field the currents in the armature wires in every revolution reach a maximum four times instead of twice, as in the two-pole field—once every time they pass a polar surface; they also become zero and change their direction four times—midway between the poles; and, if the currents are to be maintained continuously in the same directions, they must be commutated whenever these changes of direction occur. There will, therefore, be required four brushes in this case instead of two. In a six-pole field there will be six reversals and there must be six brushes, and so on; two additional brushes must be added for every additional pair of poles introduced into our field. In Fig. 50 is shown the arrangement of the brushes in a four-pole machine, the ring winding being used for illustration as being simpler for the purpose. In the bipolar (two pole) field it will be remembered that the positive brush was placed diametrically opposite the negative brush. In the four-pole field this is not so, because the north pole of the magnet is not opposite the south pole. They are at right angles to each other, and therefore in order that the brushes may have the same position relative to the field that they had before, viz., at right angles to them, they are in the four-pole field at right angles to each other also. It will be seen from Fig. 50 that the positive and negative brushes alternate, and by connecting two circuits in the proper way to these brushes we would have two independent currents. In fact, while in the four-pole field we have practically two separate bipolar magnets, we also have, when a single armature is added, practically two distinct generators or motors. Or, as is more usually the case, these two independent circuits are united into one, and we have, if everything is properly arranged and proportioned, a single machine equivalent to the output of the two considered separately. It is evident that this can be accomplished by connecting the two positive brushes (those marked +) together and to one end of the circuit, and the two negative brushes (those marked —) to the other end of the circuit, and in large multipolar generators this is usually the method employed, all of the brushes from which the current is coming (the positive brushes), being connected together and all of those into which the current is passing (the nega-

124   ELECTRIC RAILWAY MOTORS.

tive brushes) being connected together, and these are attached to the positive and negative terminals of the line circuit.

When the brushes of the same kind are thus connected together, the electromotive force of the whole armature is simply that of any of the sets of coils from one positive brush to the adjacent negative brush. In the diagram (Fig. 50), see page 126, the coils of the four quarters of the armature are in multiple arc with each other, and since there

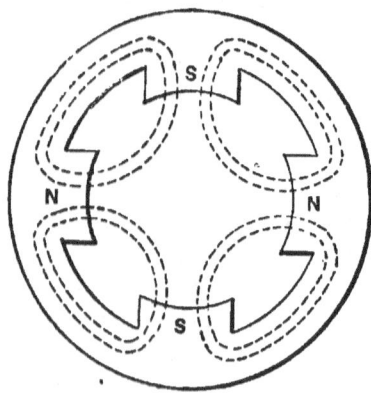

FIG. 49.

are therefore four paths of the same size for the current, the resistance offered to the passage of the current is only one-fourth what it would be if the coils of all four were in series, or if the coils on one-half the armature were in parallel with those on the other half, as would be the case were this a bipolar field and there were but two brushes.

Referring again to Fig. 50, since in a four-pole field the opposite poles are alike, the coils on the armature, diametrically opposite to each other as they sweep by these poles, will always have electromotive forces generated in them in the same direction. This being the case, we can connect the diametrically opposite coils with each other, either in series or in multiple, so that they practically form but one coil, and by bringing the ends of this coil to the proper commutator blocks, again reduce the number of brushes to two. In this case the brushes will be most conveniently located at right angles to each other, about halfway between a north pole and its two adjacent south poles, or between a south pole and its two adjacent north poles. It is evident that if opposite coils be connected in series the electromotive force generated will be doubled, while the resistance is also doubled, and if they are connected in parallel the electromotive force will be that of either coil alone, while the current will be doubled and the resistance halved. This method of connecting together the opposite coils of an armature in a multipolar field is the usual one in multipolar street car motors, for the reason that it reduces the number of brushes, which is a very desirable accomplishment, and also enables the brushes to be more conveniently located for inspection than would otherwise be possible.

We will have noticed another advantage in multipolar machines. In discussing bipolar machines it was stated that in any given case the electromotive force generated by a coil revolving in a magnetic field depended upon the number of revolutions it made per minute in that field, or, in other words, upon the rapidity with which it cut the lines of force. We have just seen that in a fourpole field the coil cuts the lines of force twice as often in each revolution as it does in a bipolar field. The armature, therefore, need revolve only *half* as fast in a four-pole field as in one of two poles to produce the same output, and thus we are enabled to make slow-speed generators. The same may be said of motors, and it will have been observed that all slow-speed motors, such as the gearless motors, are provided with multipolar fields. The philosophy of this, simply stated, is that if a given current fed to a motor having two

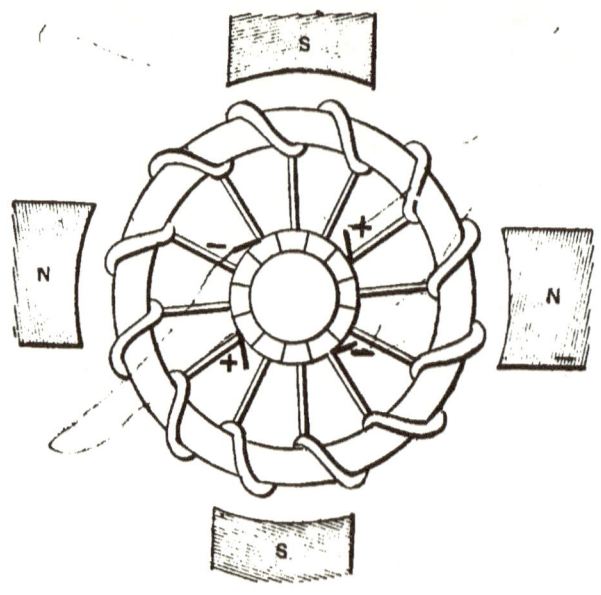

Fig. 50.

poles will give it at any specified speed a given power, the same current acting upon a motor with four poles would produce a motor practically equivalent to two machines of the bipolar type, and as the power of a motor is also dependent upon its speed, a four-pole motor will produce the same power at half the speed of a similar motor with only two poles.

Thus far we have said nothing as to how we get the magnetism in our field magnets. We have assumed, however, that they are electromagnets, viz., that the field magnets are made of soft iron and have no magnetism of their own, but are converted into magnets by passing electric currents through coils of insulated wire properly wound around them, and this is now the universal practice in all but very small generators. Permanent magnets made of very hard tempered steel might be used, however, but as a permanent magnet can never be made so strong as an electromagnet of the same size, and for other equally good reasons which need not be mentioned here, the latter are preferred. The earlier dynamos were, however, frequently constructed with permanent magnets. Such machines are properly called *magnetoelectric* machines. Examples of magnetoelectric machines are found in the apparatus employed in the telephone fixtures for ringing up the exchange. In the call box will be found a strong permanent magnet between whose poles there revolves a small iron bobbin (armature) wound with fine wire. As this is rotated by turning the crank it generates alternating currents which pass over the line and ring the bells. It would be a simple thing to attach a commutator to the armature which would convert the alternating currents into direct currents, as already described, but the telephone call bell has been adapted to alternating currents, so that the complication of commutators is not necessary.

Another method of obtaining our magnetism is to construct our fields of soft iron and wind them with coils, and connect these coils with some independent source of electricity, such as a large battery or another dynamo. When the current flows through the coils, of course the fields become highly magnetized, as we have seen. A machine whose fields are thus excited is called a *separately excited* dynamo. This method has some advantages, chief of which is that the exciting current, coming from a separate source, is entirely independent of the fluctuations of current in the trolley circuit, which would, if current from the latter were added, produce similar variations in the magnetism of the fields, which would in turn still further complicate matters. It has the disadvantage, however, of requiring a separate machine for this purpose, which in very small plants would be scarcely warranted by the attendant advantages. In large electric power stations it is quite customary to find a small machine used solely for this purpose, its current being employed to excite the fields of all generators in the plant.

CHAPTER XIV.

THE DYNAMO-ELECTRIC PRINCIPLE.

THERE is still a third method, and this is the most usual one, viz., to make the dynamo excite its own fields by causing either all or a portion of the current generated by the armature to pass around the field magnet coils. But the question naturally arises, how are we to start such a machine into action? When the armature stops, the current stops, and the magnetism of the fields disappears. If we start the machine from rest, there being no magnetism in the fields, there will be no lines of force for the armature wires to cut; the armature will therefore generate no current, and our provision for utilizing that current to excite our field will be useless. So reasoned the early builders of electrical machines, and it was thought necessary for a long time to separately excite the fields, at least until the machine got into action. It was therefore a very important discovery that such was not necessary. It seems that all iron, however soft, has a *little* magnetism, which it either derives from the earth's magnetism or retains from previous magnetization (residual magnetism).

This may be very slight, but it is sufficient so that when the armature is revolved before the pole pieces it generates a very slight current. No matter how insignificant this current may be, as it passes around the field magnets it adds somewhat to their magnetism. This increased magnetism of the fields enables the armature to generate a little stronger current than before, and this produces more magnetism and that more current, so that by the continued reaction between the field and armature the machine "builds itself up," as the phrase is, until the magnets have arrived at full strength, and the armature is putting out current to its maximum capacity.

This "building up" of the machine from practically nothing to its maximum output without outside help except from the power necessary to drive the armature, seeming at first sight to be very much like the attempt to lift one's self over the fence by one's boot straps, is known as the "dynamo-electric" principle. These "self-exciting" machines, which are by far the most numerous of those employed to-day, are the true "dynamo-electric" machines.

SERIES, SHUNT AND COMPOUND WINDING.

The dynamo-electric principle gives rise to three distinct types of machines. In the first type all of the current from the armature passes around the field coils, as in Fig. 51, before it goes out to the exterior circuit. Since the field coils are in *series* with the armature, a machine so wound is called a *series* dynamo or motor, as the case may be. It is evident that if resistances are placed in the outer circuit the amount of current that will flow around the coils will be lessened. This will lessen the strength of the field magnets, and this will still further lessen the amount of current that the armature can give; and if the exterior circuit

SERIES, SHUNT AND COMPOUND WINDING. 131

becomes broken so that *no* current can flow, of course the field magnets, no longer having any current in their coils, lose their magnetism, and the armature ceases entirely to generate current. If the break in the outer circuit be now closed again, the dynamo will gradually build itself up, as before described under the heading "Dynamo-Electric Principle," until it again attains its full strength, but this will take an appreciable time. A series dynamo, of course, could not be used on a multiple

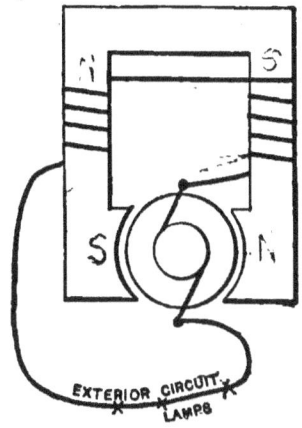

FIG. 51.

arc street railway or on the usual incandescent circuit, for on either of these circuits we want to have at our command the full strength of the current the moment we start the first car or turn on the first light. We cannot wait for the dynamos to build themselves up. There are other objections also to series dynamos for these purposes, but for other purposes they have their use.

In the second type (Fig. 52) we only use a portion of the current generated by the dynamos to excite the fields. From the brushes there are two circuits, between which the current divides; one of these is the line circuit, which is of heavy wire sufficient to carry all the current required in the exterior circuit, and the other is a thin wire of great length, and therefore of high resistance, which is wound in many turns around the field magnet cores. Since this latter wire is of high resistance, but little current passes through it; but as it passes many times around the magnet there

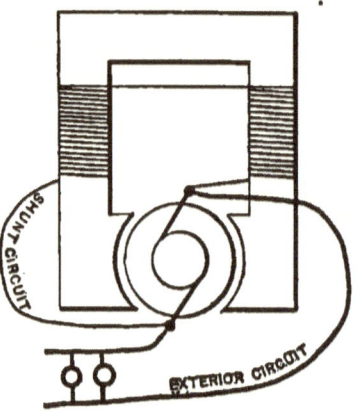

FIG. 52.

are sufficient ampere turns with the small current for our purpose. This latter wire, which forms in this case the magnet coils, is in multiple arc, or in parallel, or "*in shunt*," with the exterior circuit. This type is therefore called the "shunt dynamo."

There is a law in the flow of electric currents to which we have before referred, that when several paths are open to the current the latter will divide itself among them according to the relative conductivities of those paths. In Fig. 52 the cur-

rent has two paths open to it—one through the
fine wire of high resistance, which forms the
magnet coils, and the other through the large wire,
which forms the external circuit. The shunt circuit, as will be observed, is always closed, so that
no matter whether the external circuit be closed
or not a current will always be passing through
the field coils, and the full magnetism which they
are capable of imparting to the magnet is always
maintained. The shunt dynamo is therefore

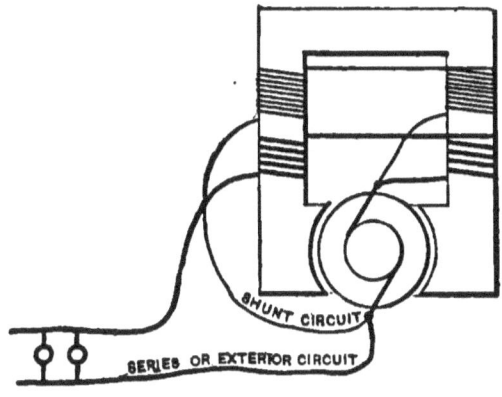

FIG. 53.

peculiarly fitted for multiple arc circuits (street
railway and incandescent lamp circuits), because, as
we know, multiple arc circuits are always open
when there are no translating devices (lamps,
cars, etc.) operating. But if the magnetism of
the fields be maintained, the full current is always
available the moment it is wanted. But the
exterior circuit is one of variable conductivity.
The resistance between the trolley wire and the
ground is only half as great when two cars are
running as when only one is in operation. The

relative conductivity of the two paths therefore changes with the number of cars operated, being greatest in the outside circuit when there are many cars.

Now supposing that with the speed at which our armature is driven, and the magnetism which the ampere turns of our shunt coils is capable of producing, our dynamo is capable of generating an electromotive force of just five hundred volts when but one car is in operation. If a second car be started up, the conductivity of the exterior cir-

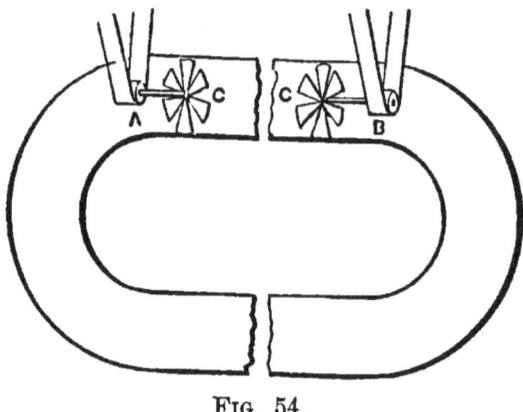

FIG. 54.

cuit will have been increased, and it will take a relatively larger portion of the total current generated by the armature. This will leave less to go around the magnet coils, and the magnets become weaker. With weaker fields the electromotive force generated by the same speed of armature becomes less, and we will no longer have a current of five hundred volts, but something less than that. This loss of electromotive force is what is technically termed "drop," and while in very short lines with but a car or two operating it may not amount to much, it becomes very serious in long lines with many cars or lights to feed. To correct this fault we sometimes have recourse to the third method of winding, which is known as *compound* winding, and the machines so wound are known as compound dynamos. The compound winding (Fig. 53) is a combination of the series and shunt windings. As shown in the diagram the winding consists of two coils, one consisting of a few turns of coarse wire in series with the armature, through which all the current which goes to the exterior circuit passes, and the other consisting of the fine wire coils in parallel or in shunt to the latter. Now if we reduce the resistance of the exterior circuit by putting on additional cars, while we weaken the magnetizing effect of the fine shunt coils, as in the last case, the additional current which is diverted to the exterior circuit, which has to pass through the series coils, compensates for the other loss. By putting on fewer or more turns of the series coils we may exactly compensate for any loss of electromotive force that might be occasioned in a shunt machine by the addition of cars or length of line, or by putting on more turns than is needed, for exact compensation may even cause the electromotive force to rise as the load on the dynamo increases. A machine wound to produce the latter effect is said to be "*overcompounded.*"

THE REVERSIBILITY OF THE DYNAMO.

It will have been observed that heretofore we have referred to the dynamo and to the motor as though they were synonymous terms. We have done this not indiscriminately, but as the one or the other served the purpose of illustration the better. But as a matter of fact the dynamo and the motor *are* one and the same machine.

Supposing we have in a pipe, C, two fans exactly alike (Fig. 54), each furnished with a pulley by which it may be belted to a line shaft or to other machinery. If we drive A rapidly, it will force a current of air through the pipe C from A to B, and this current of air will cause B to revolve like a windmill. If, on the other hand, we drive B by its pulley, the current of air which it will produce will drive A, and if the current be strong enough the latter, acting as a motor, may drive other machinery to which it may be belted. In the first case we have driven the fan A by belting it to a driving pulley, and it has become a generator of a current of air which, upon being conducted by the pipe C to the similar fan B, has caused the latter to revolve as a motor, rendering it capable of driving other machinery through its pulley and belt. In the second case B as a generator drives A as a motor.

It is evident that these two fans are perfectly reversible as regards their functions. By driving either one it becomes a generator of air current capable of driving the other one as a motor. That is to say, if we apply mechanical energy to either, it gives out wind energy, if we may use such an expression, and if we apply to it wind energy, it will give out mechanical energy.

So it is exactly with the dynamo-electric machine. If we drive the armature by steam or other power, the machine will generate an electromotive force—electrical energy; and if we apply electrical energy to its armature, the latter will revolve and give out mechanical energy. If two exactly similar machines have their brushes connected by electrical conductors, they will behave toward each other exactly as do the fans. Since they are exactly alike, it is immaterial which we shall employ as a generator and which as a motor. If the armature of one is driven, it will give rise to an electric pressure corresponding to the pressure of air in C, which by giving rise to a current will cause the other to revolve as a motor, the general rule being that any machine that will make a good generator will also make a good motor, and *vice versa.*

It will be seen from the above that the electric motor bears the same relation to the generator as the driven pulley on one shaft does to the driving pulley on another, and that the electric current by which the energy is conveyed from the generator to the motor performs the same office exactly as the belt does which connects the driving pulley with the driven pulley. It is perfectly clear that either of two lines of shafting may be used as the driving shaft by connecting it with the steam engine, and if the pulleys which are belted together on the two shafts are of the same size, it will make no difference in the operation of the machinery to which shaft it is belted, but in mechanical operations it is often desirable to give the driven shaft a different speed from that of the line shaft, and for that reason the pulleys on the two shafts would be given different diameters to adapt them to the required conditions. So in electrical machinery the motor may differ radically in *appearance,* and also differ somewhat in minor details, from the generator, to better adapt it to the particular work it has in hand. Thus in the street car motor compactness is a prime requisite, and it is allowable to sacrifice some of the requisites of a good machine in order that this one feature may predominate. But motors will not differ more in appearance from generators than they do from each other. While the construction of the two machines—the dynamo and motor—may be and frequently is the same, the theory upon which they operate is entirely different, the one being the reverse of the other. The same drawings are, however, entirely applicable to the explanation of the motor.

CHAPTER XV.

THE ELECTRIC MOTOR.

WE have seen (Fig. 10, see page 60) that if an electric current is passed through a coil of wire it sets up lines of force which have a definite direction within the coil and give the coil a distinct polarity. Now if this coil while traversed by a current of electricity be brought into a magnetic field, viz., be placed between the poles of a magnet, it will tend to take up such a position that the lines of force generated within its own coils shall have the same direction as those of the magnetic field in which it is placed. That is, it will tend to place its own axis directly on the line joining the north and south poles of the field magnet, with the south pole of the coil facing the north pole of the magnet, and the north pole of the coil facing the south pole of the magnet. In this position the coil presents the least obstruction to the passage of the lines of force between the two poles of the magnet, because it presents the greatest area for their passage, and in fact assists the passage of these lines by the magnetomotive force which its own current adds to that of the magnet between whose poles it is placed. When the coil is in the position shown in Fig. 23 (see page 84), with a current traversing it in such a direction as to assist the lines of force across from N to S, we have this condition fulfilled, and this is the position which any coil traversed by a current will take up, if free to move, when placed between the poles of a magnet. The usual way of expressing this fact is to say that a closed electric circuit when placed in a magnetic field tends to take up such a position as to enable it to embrace the greatest possible number of lines of force. Clearly this condition is best fulfilled when the plane of the coil is at right angles to the lines of force, as in Fig. 23, and least fulfilled when it is in the position shown in Fig. 22 (see page 83), because in the latter the plane of the coil is parallel to the lines of force passing from N to S, and it can embrace no lines of force unless by some means they may be diverted from a direct line so as to thread themselves through the loop in curved lines. This they are induced to do by the current in the coil, which sucks them in, as it were, either from the upper side or the lower side of the coil, according to the direction of the current in the coil. With the current flowing—as indicated by the arrows in Fig. 22—in the direction of the hands of a clock as we look down upon the coil, the upper surface will have a south polarity—that is, its own lines of force will enter the coil from that side and emerge from the lower side, which will have north polarity. In like manner some of the lines of force which extend in straight lines from N to S will be bent out of the direct line so as to thread themselves through the loop in the same way. Other lines will crowd through, and in doing so, and in trying at the same time to straighten themselves out again, will pry the loop around from its present position in a direction contrary to that indicated by the arrow. As the angle through which the loop is thus turned increases it permits still more lines to thread it, and these add their prying effort, until the plane of the loop is at right angles to the lines, which permits the lines to pass through without bending, and therefore without further tendency to rotate the coil.

We may note right here one peculiar fact: We found that when we revolved the coil between the poles $N\,S$ from left to right in the direction indicated by the arrow it produced a current in the direction $A\ B\ G\ C\ D\ A$. Now, if we pass a current through the coil in the same direction as it took when the coil was mechanically revolved, viz., $D\ A\ B\ G\ C\ B$, it tends to cause the coil to revolve in the *opposite* direction. That is to say, the current which is generated by revolving a coil in a magnetic field is in such a direction as would cause the same coil to revolve in the opposite direction: the action of the electrical current generated in a coil is to directly oppose that of the motion which produced it. This is a fundamental law of electrics, and explains why, as the current increases in a circuit, it requires more force to drive the armature of the dynamo. It is because the larger the current traversing the coils of the armature, the greater the effort on the part of that current to revolve the armature in the opposite direction, and it is the energy that is absorbed in overcoming this tendency that reappears in the armature coils as electricity.

But to go back a little ways. If with the coil in the position shown in Fig. 22 we should pass the current in the opposite direction to that indicated by the arrows, the lines would thread through the loops from the under side and come out on the upper side, thus prying the loop over in the opposite direction. If at the same time that we change the direction of the current in the loop we also change the direction of the lines of force in our field by making the right-hand field north and the left-hand field south, we see that the lines will enter the coil on the right hand from below and emerge on the left hand from above, and in the effort to straighten themselves out will again tend to pry the loop in the same direction as in the first case. We therefore see that reversing both the magnetism of our fields and the direction of the current in the armature has no effect upon the direction of rotation of the armature. The direction of rotation will be changed, however, if either of them alone be changed.

If we place a single coil of wire traversed by a current in a magnetic field it will tend to revolve about its axis until the plane of the coil is at right angles to the lines of force of the field in which it is placed, and the direction of its rotation will be determined by the direction of the current which flows through the coil. No matter how strong the current or how powerful the field, the coil will not tend to revolve further than 90° from the direction of the lines of force. This is entirely similar to the action of the compass needle when a current-carrying wire is placed over it. We remember that under these conditions the needle was deviated in one direction or the other according to the direction of the current in the wire, and tended to take up a position at right angles to the wire. It would be a more general statement, but equally true, that the wire had an equal tendency to place itself at right angles to the needle. The action was more apparent in the needle, however, because that was readily movable, while the coil was not.

But suppose we have two coils of wire with their planes say at right angles to each other, as in Figs. 30, 32 and 33. If the current be diverted from that coil which has already been revolved to a position at right angles to the lines of force of the field, to the other coil, which is now parallel to those lines, as may be automatically accomplished by a commutator of four parts, the rotation will be continued in the same direction, and we have at once an elementary electric motor. Or if we have but a single coil, as in the first case, whose ends terminate in a two-part commutator, and the direction of the current in the coil be reversed when it has reached a position at right angles to the field, its own lines of force will be reversed in direction and cause the field lines to seek a passage by a circuitous route from the opposite side of the coil, and these, as before stated, in their endeavor to straighten themselves out, will pry the coil over still further and cause it to make another half revolution.

In an electric motor advantage is taken of both of these actions where there are many coils at various angles to each other, and as each coil comes to the position where the threading of the lines of force through it in one direction exerts no further tendency to cause it to rotate, the current in that coil is reversed in direction so as to cause the lines to enter from the opposite side and continue to exert an effort at rotation. Thus by changing the direction of the current in the coil at the proper time twice in each revolution a single coil may be kept in continuous rotation. But the effort which will be exerted will vary widely with different positions of the coil with respect to the field. Twice in every revolution, viz., when the coil is at right angles to the field the effort will be nothing, and twice, viz., when the coil is parallel to the field, it will be a maximum. We find that these positions correspond with the positions of minimum and maximum activities of the coil when used to generate current. We remember that in describing the dynamo with two coils at right angles to each other it was stated that the current would be more uniform, for in that case when one coil was in its neutral position and cutting no lines of force, the other one would be in the position where it would be cutting the lines of force at a maximum rate. So with the motor. If we have two coils at right angles to each other, one of these will be exerting its greatest effort at rotation, while the other one is exerting none. Thus by increasing the number of coils at small angles with each other the effort to turn the armature will not only be increased, but become more uniform. The effort exerted by the armature to revolve is termed its "torque," and is dependent upon the number of lines of force that can be induced to thread themselves through the coil, and this is dependent, as we know, upon the current. The "torque" of a motor, or the effort which it exerts to turn against a resistance, is therefore said to be dependent upon the amount of current passing through the armature coils. It is also, of course, dependent upon the strength of the field, but if that be constant it is proportional always to the current.

CHAPTER XVI.

THE ELECTRIC MOTOR.—(*Continued.*)

ANOTHER way of explaining the action of an electric motor, which is simpler, but in most cases not so correct, is the following : Let us suppose for the moment that there is but a single coil on our armature (Figs. 55 and 56). If in looking at the commutator end of the armature the current passes through the coil as it passes over the end of the armature coil in the direction of the arrow (Fig. 55), which would be clockwise in the coil if we look at the coil from the north pole of the magnet, and anti-clockwise if viewed from the south pole, it will make of the armature core an electromagnet whose south pole is near the north pole and whose north pole is near the south pole of the field magnets. Since unlike poles attract each other, the armature will tend to revolve until these unlike poles are as near together as they can get, or until the axis joining the two poles of the armature is in a straight line with or parallel to the axis joining the field magnet poles. If nothing more were done, the armature would simply oscillate back and forth a few times on either side of this line, and finally come to rest in the position stated, just as a compass needle does when a magnetic pole is brought near it, and the attractive action of the unlike poles upon each other would oppose any effort to move the armature in either direction. But we have seen that when these four poles are in line the coil on the armature is in its neutral position, viz., in that position where in the dynamo the electromotive force generated in the coils changes direction, and where by means of the commutator it is rectified for the exterior circuit. If, therefore, a direct current enter the armature coil through the branches and commutator, its direction in the coil will be reversed at this point. Figs. 55 and 56 represent the coil just before and after it has occupied this neutral position, and the arrows show the directions of the currents under these conditions. It will be

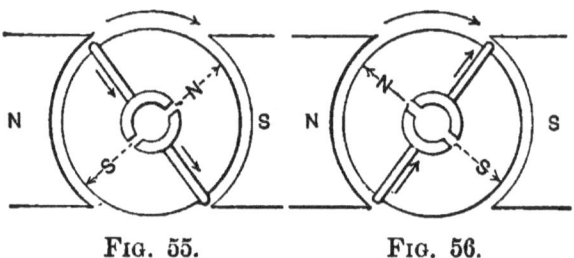

FIG. 55.   FIG. 56.

noted that just at the moment when the south and north poles of the armature have reached the point beyond which the mutual *attractions* between them and the north and south poles of the field magnets would no longer tend to cause the armature to revolve, the direction of the current is reversed by the commutator, and what were the south and north poles of the armature become the north and south poles, and like poles of the machine are brought into proximity and *repulsion* occurs and rotation is continued. Thus by properly arranging the coils and commutating the directions of their currents a continuous effort, first of attraction and then of repulsion, is exerted upon the armature, which causes it to keep in continuous rotation and enables it to do more or less work as this pull and push is large or small.

TORQUE.

Now we know that the strength of a magnet—viz., the ability which it evinces either to attract an unlike pole or to repel a like pole—depends, other things being equal, upon the number of ampere turns upon the magnet. In the case of the motor or dynamo the number of turns on both the field magnet and the armature is fixed, so that the only way in which we can vary the strength of these magnets is by increasing the current (remembering that the ampere turns $=$ number of turns of the wire $\times$ current in amperes). We can therefore increase the turning effort of the armature by increasing the current and thus increasing the strength of the poles. This effort to revolve which the armature is able to exert is technically termed its "torque." The torque, therefore, is dependent, among other things, upon the current, being greater or less as the current in the armature is greater or less.

But everyone knows that in turning a capstan, or in winding up a heavy weight attached to a rope on a windlass, it can be moved much more easily if the crank arm or lever is long than if it is short. In fact, to give a concrete example, if our windlass drum have a diameter of 1 foot and the load on the rope be 100 pounds, this can be exactly balanced by hanging on the end of the crank arm a weight of 10 pounds if that crank arm be 10 feet long, and 11 pounds so placed would draw the heavier weight of 100 pounds up. In this case the torque due to the force of 10 pounds acting on the end of a lever arm 10 feet long is exactly equal to the torque due to a weight of 100 pounds acting at the end of a lever arm but 1 foot long. Thus we see that while we can increase the pull and push on the armature by increasing the current, we can increase the effect of this pull or push, or, in other words, make a machine which will have still greater torque, by increasing the length of the lever arm upon which the torque due to current acts. As an example parallel to the one cited of the windlass, an armature ten feet in diameter would, with the same current, have ten times the torque that one but a foot in diameter would have. But in street car motors our space is limited and we cannot far increase the torque of our motor in this way. Our armature must necessarily be of small diameter, so that our leverage is small. Nor can we increase our current indefinitely, so other means must be resorted to to give the motor sufficient torque to start a loaded car from rest or propel it up a steep grade.

It is a law of mechanics that what is termed "work" is equal to the product of the force (torque) into the space described by it in its direction while overcoming the resistance, and "horse power" is the rate at which this work is done. Thus if we have an armature of small diameter and small torque, it may be able to do considerable work at a rapid rate if with this small torque it be caused to revolve very fast. If this were attached directly to the axle of a car, its torque would not be sufficient to start it, but if we geared it down through one or more gears so that for every revolution of the axle the motor would in the same time have made say fifty revolutions, the torque on the car axle would be magnified fifty times, and the car would easily start or go up hill. It must not be imagined that by gearing down a motor is enabled to do more work in a given time, or that its horse power is in any way increased thereby, for by the definition of work it is the product of the force into the space through which it acts. If we exert a small force through a very long distance in a given time, as is

GEARLESS MOTOR.

Fig. 57.

the case with the rapidly running armature, the same horse power is expended as if we exerted fifty times as much force through a space one-fiftieth as long.

Thus in street car motors with the small armatures and bipolar fields the necessary torque on the car axle is gained at the expense of speed by gearing down. At first all street car motors were double geared, but the wear of these gears and the loss of power occasioned by them indicated the desirability of doing away with them either partially or wholly. This could only be done by increasing in some way the torque on the armature. Neither the diameter of the armature nor the current used could be much increased, so resort was had to the multipolar fields. It is very clear that with a four-pole field each of the fields may be made to exert the same pulling or pushing force on the armature with the same current as is exerted by each of the bipolar fields before referred to. With four poles, therefore, the armature will have double the torque, and may therefore run at half the speed with the same mechanical advantage. This would enable us to do away with one of the two reduction gears of which we have spoken, and if we had six poles and could slightly increase the diameter of our armature, or slightly increase the amount of current, or both, we could put the armature directly on the axle and do away with both gears and have the same axle torque as we originally had with our double gears. We would have then a single reduction and gearless motor respectively. It was argued that whatever excess of current might be required for the gearless motor, for instance, would be more than compensated for by the saving of the loss in the gears. As an illustration of the gearless motor reference is made to the subjoined cut of the Short gearless motor. It will be observed that its field frame is triangular in section and has but three pole pieces. It is, however, a six-pole machine, since the fields are so wound as to produce consequent poles in the centers of each of the three parts of the frame connecting the three pole pieces together.

CHAPTER XVII.

THE LINE OF COMMUTATION.

HERETOFORE in describing the action of the dynamo we have represented the two neutral positions of the wire as being on a line at right angles to the axis of the pole pieces (see Figs. 26 and 28, see pages 87 and 89). This would only be the case, however, when there was no current in the armature. Let us examine what would happen in a motor armature when a current is passing in its coils. If it be a closed coil armature, such as is now almost universally used on street cars, the current passing through the armature coils will tend to make north and south poles on the armature, which will be at the ends of a diameter at right angles to that which joins the north and south poles induced in the armature by the magnetism of the fields. Since under the circumstances there cannot be more than two poles in the armature, there will be a compromise between the north and south poles produced by the armature coils and the corresponding north and south poles induced by the field magnets. Should these two sets of poles be of exactly equal strength, the compromise poles will be halfway between the two, and the line of commutation (the line which joins the positions where the individual coil is generating no electromotive force) will be at right angles to the line joining the compromise poles, and the brushes must be moved to this new position, else they will pass from one commutator segment to the next while there is a difference of potential between them, which will cause sparking.

If the induced poles are relatively stronger than the poles produced by the armature current, which would be the case if the armature current is weak compared with the field strength, the compromise poles would approach more nearly to the induced poles, and if the field magnets were very overpowering in their strength, the compromise poles would nearly coincide with the induced poles, and the line of commutation would nearly coincide with that which we have provisionally given it, viz., at right angles to a line joining the field magnet poles. But as a motor is called upon to do more or less work, the current admitted to its coils must be increased or decreased. This results, of course, in strengthening or weakening the resulting poles. The position of the compromise poles will therefore be constantly changing, which means, of course, that the line of commutation is constantly shifting. It has already been explained that to prevent sparking the brushes must bear on the armature on the line of commutation, so the brushes must be shifted as that changes. By making the armature poles weak relative to the field, considerable change of current in the armature may occur without much shifting of the neutral line, and under these circumstances it may operate through quite a range of work with fixed brushes without sparking. It is because of an abnormal rush of current through the armature on starting up a car, resulting in an abnormal shifting of the neutral line, that the sparking in street car motors occurs. Since the street car motor is called upon to operate through an exceedingly wide range of work, the field magnets are made the dominating ones so as to keep the neutral line very nearly stationary throughout this range. In series motors, however, if neither armature nor field is near saturation, the neutral line will not change its position.

COUNTER ELECTROMOTIVE FORCE.

Considerable space was devoted to explaining and emphasizing the fact that if a closed loop of wire be revolved around its axis between the poles of a magnet, or, in other words, is revolved in a magnetic field, so that its rate of cutting the lines of force is constantly changing, it will generate an electromotive force.

It is perfectly apparent that when we operate a motor by passing current through it from one brush to the other we are fulfilling all of these conditions, and the coils of the motor armature must also develop an electromotive force. It makes no difference whether an armature is driven by electricity or by steam power, if its coils revolve in a magnetic field, there will be generated in them an electromotive force. In the case of the motor, however, this electromotive force is in the opposite direction to that of the current which operates the motor, and is therefore called the "*counter*" electromotive force. It follows exactly the same laws as govern the generation of electromotive forces in dynamos, and will be the higher the greater the speed of the armature. A motor running at high speed, as it will if allowed to when it is running without load, will generate a counter electromotive force almost equal to that of the current by which it is operated, and will therefore take but very little current. The current that a motor will take under any conditions is that which is due to the difference between the direct electromotive force of the current and the counter electromotive force of the motor itself. The counter electromotive force of a motor at any speed is exactly equal to that which the same motor would generate if driven as a dynamo at the same speed. If, therefore, a motor is starting from rest, it has, at first, no counter electromotive force. The only obstacle to the flow of current through the armature would be the resistance within the machine itself, which, being exceedingly small, would, according to Ohm's law, permit an enormous current to flow through its coils, which would inevitably result in a burn out if it were not checked. This is the reason why a rheostat is almost universally used in starting. It opposes at first a great artificial resistance, and this is decreased as speed is gained and counter electromotive force is generated, when all artificial resistances may be removed with impunity.

This back electromotive force of a motor has its exact counterpart in the resistance which the armature of a dynamo offers to the driving engine. We all know, or can find out for ourselves, that when a dynamo is generating no current, as, for instance, when the armature is not in motion, we can readily turn the armature with the hand, and yet when it is working to its utmost capacity it may take hundreds of horse power to keep it in rotation. This resistance to the mechanical effort of the steam engine is really the measure of the work the dynamo is performing; so the electrical resistance opposed by the counter electromotive force of the motor is a measure of the work the motor is doing.

Professor Silvanus Thompson, in his great work on "Dynamo-Electric Machinery," cites an example which very well shows how the current that a motor will take decreases with the speed of its armature. He used a small Immisch motor with separately excited fields, and connected it up with a primary battery and amperemeter. At different speeds the following figures were obtained:

Speed Revs. per Minute.	Current Amperes.	Speed Revs. per Minute.	Current Amperes.
0	20	160	7.8
50	16.2	180	6.1
100	12.2	195	5.1

Thus at its maximum speed it took only about one-fourth of the current that it took when the armature was held at rest. In this case 5.1 amperes were required to overcome the friction of the armature. Had the friction been less the armature would have revolved still faster and finally come to constant speed with less current than 5.1 amperes.

In the earlier days of the electric motor this counter electromotive force was a bugbear and thought to be an objectionable feature, and attempts were made to construct motors from which it would be eliminated. But we now know that the existence of this counter electromotive force is of the utmost importance, and that upon it depends the degree to which any given motor enables us to utilize electric energy that is supplied to it in the form of an electric current. "In fact," says Professor Thompson, "this counter electromotive force is an absolute and necessary factor in the power of the motor, just as much as the velocity to which (other things being equal) it is proportional."

As will be seen from the figures given by Professor Thompson and the subsequent remarks, the amount of current that a motor will take is only that which is absolutely necessary to do the work which it is called upon to do. That is to say, if it has no other work to do than to overcome its own friction, its armature will automatically attain such a speed as to generate a counter electromotive force, or "back" pressure, such that only sufficient *effective* electromotive force remains to force through the motor sufficient current to move the armature against this resistance. If now an additional load is thrown on the motor, the speed of its armature will be at once retarded, less counter electromotive force will be generated, and consequently more current will pass, and the motor at once adjusts itself to this new load.

CHAPTER XVIII.

COUNTER ELECTROMOTIVE FORCE AND SPEED REGULATION.

But this counter electromotive force must not be confounded with dead resistance. It has in some respects the same effect as resistance, but differs from the latter in very important particulars. It has been stated that counter electromotive force cuts down the current. Resistances placed in circuit will do the same thing exactly, but the cutting down of current by means of the counter electromotive force, which tends to produce a current in the opposite direction, is merely a problem in subtraction—the taking of a lesser quantity from a greater, and the utilization of the remainder. No energy whatever is consumed when the current is reduced in this way, but when it is reduced by the use of resistances the electromotive force which disappears is employed in overcoming these resistances, and the energy thus expended appears as heat. In electric lighting, welding, etc., this energy is usefully expended, for by designedly localizing the resistance we localize the heat to such an extent as to produce the desired results of high temperatures. But in most other applications of electricity it is not heat that we want, but mechanical energy. Every bit of heat that is generated in our circuit or in our translating devices, therefore, represents so much energy uselessly employed, and constitutes a drain upon our source of energy for which we have to pay as much as we pay for the energy which serves a useful purpose. The heat produced in an electric circuit corresponds very closely to the water lost in a leaky water main along the route. If the pipe be very leaky, half the water that is pumped into it at the pumping station may leak out, and while it does the consumer at the other end of the line no good, costs at the pumping station just as much as the other half does which he can make use of. Where an electric motor heats up badly, or resistances are introduced into the circuit near the motor in order to cut down the current to the required amount, they constitute large leaks which, in the water analogy, would correspond to the diversion of a portion of a waterfall from a water wheel so as to let that portion diverted run to waste, or where the water is conveyed to a water motor in a pipe, to the opening of a large faucet just above the water motor, so that a portion only of the water that was delivered at that point would go to the motor, the remainder being allowed to run out upon the ground.

But it is often necessary with electric motors to resort to these wasteful methods of regulation. As everyone knows, it requires the expenditure of considerable energy to set in motion any body which is at rest, and it also requires the expenditure of this energy for a considerable time before that body—especially if it be a heavy body—can be set into very rapid motion. Everyone also knows that when the moving body has once attained a certain speed this same speed can be maintained with the expenditure of but a very

COUNTER ELECTROMOTIVE FORCE. 159

small fraction of the energy required to bring it up to that speed. The same is true, in a reverse order, of bringing a moving body to rest. This property, which is characteristic of all matter, by which it resists any change as regards motion or rest, is termed *inertia*, and the heavier the body is the longer must a given force act upon it to start it from a state of rest and bring it to a given speed, or to check it from a given speed and bring it to rest. We therefore see why in electric motors it is necessary in starting them to reduce the current by what have been termed dead or hurtful resistances, for when the armature is at rest there is of course no counter electromotive force to oppose the current. The only resistance opposed to its flow at this instant is that which is offered by the coils on the armature (and fields in the case of series winding), and this has been made by the builder as small as possible purposely to prevent the loss of energy which resistances necessarily involve. If, therefore, the current were turned on full from a five hundred volt circuit, the momentary rush would be so great as to burn the motor up at once.

To realize how quickly the heating effects increase with the current we need only to remember the law that the heat thus produced is proportional to the *square* of the current. It would be bad enough under these circumstances if double and treble the current produced only double and treble the heat, for a motor in starting may take ten times the amount of current that it would require when doing the full work for which it was designed, but when we remember that the heat will be increased to *four* times and *nine* times for double and treble the current, and *one hundred* times for ten times the current, the seriousness of such a situation will be realized at once.

Some means must therefore be resorted to to prevent this enormous rush at starting, and the one usually employed is dead resistance. A large resistance, sufficient to cut the current down to the safe amount, is first introduced. As the armature starts to revolve it immediately begins to generate a slight counter electromotive force. By reason of this the full amount of the original resistance is no longer required to maintain the same current, and a smaller resistance may be substituted. As the speed of the armature in-

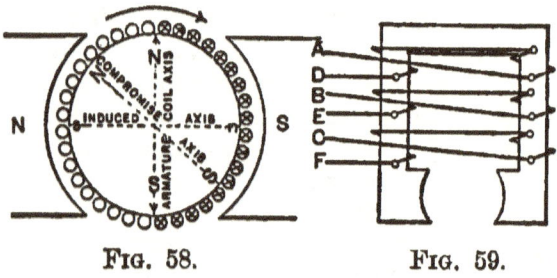

FIG. 58.   FIG. 59.

creases so does the counter electromotive force, and in like manner does the necessity for dead resistance decrease. In fact, with the speed of the armature the counter electromotive force gradually usurps the functions of the resistance until the speed has reached that at which the motor was designed to run, when it will have supplanted it entirely.

To facilitate the introduction into the motor circuit of resistances of various values as required, such resistances are usually grouped together and their terminals so connected that any one or more of them may be thrown into circuit by the movement of a lever. Such an arrangement is called a *rheostat*, and such, in fact, is the arrangement on the platforms of electric cars to which the controlling lever is attached by which the motorman controls his car.

The desire to avoid as much as possible the losses necessarily involved in the "rheostat regulation" of current as well as the requirements for speed regulation, so essential in street cars, has resulted in numerous other arrangements, which will be referred to after we have considered what the requirements of speed regulation involve.

THE REQUIREMENTS OF SPEED REGULATION.

In discussing this question it will be understood, of course, that we are not speaking of how to build motors suitable for various speeds, but how to regulate the speed of motors already built or in use. We must, therefore, take the motor as we find it, viz., with certain elements, such as the number of turns of wire on the armature, which, if changed, would modify the speed, fixed. We cannot do better on this subject than to quote from Crosby and Bell's most excellent treatise on "The Electric Railway," which those wishing to go more deeply into the subject of electric railways than it is intended to do in this volume should certainly read:

"The other quantities, changes in which are connected with changes in speed, are (1) strength of field, (2) the rate of work, *i. e.*, the quantity of work done in a given time. As has been already shown, the rate of work is measured by the product of (3) the current and (4) the counter electromotive force. The current, however, is readily expressed in terms of the counter electromotive force, the resistance of the machine and the applied electromotive force, since it is always such a current as will flow over the given resistance under a pressure equal to the difference between the two opposing pressures—that applied or impressed by the dynamo through the line, and that generated by the motor armature itself. (5) As seen from the above, change in the applied E. M. F. is also connected with change in the speed. The quantities (1), (2), (3), (4), (5) are interdependent, but are separately mentioned, since convenience requires reference first to one, then to another.

" Of these five quantities perhaps that which in practice is most constant is strength of field. As has been shown, the efficiency of a motor depends upon the relation of the counter E. M. F. to the applied E. M. F. High efficiency or relatively high counter E. M. F. is desirable at any and all speeds. But high counter E. M. F. goes with a large product of the three factors : (*a*) number of armature loops, (*b*) speed of rotation and (*c*) strength of field. As noted (*a*), the number of loops is fixed, (*b*) the speed is limited by practical requirements, hence (*c*) great strength of field is constantly desirable. It is therefore good practice so to wind a street railway motor by putting a relatively large number of turns around its magnet that a maximum strength of field is attained, even when the current flowing is small as compared with the maximum current.

" This is equivalent to saying that the magnetization given by a relatively small current is yet sufficient to saturate or nearly saturate the iron of the magnetic circuit. If, however, the field be kept below saturation and be varied in strength, this variation may be used to accomplish a certain degree of speed regulation. Let us suppose a car moving on a level (or on a uniform grade) and at such a speed as to produce a counter E. M. F. of 400, the applied E. M. F. being 500. For convenience assume the internal resistance of the armature circuit to be 10 ohms. Then the current flowing will be

$$\frac{500-400}{10} = \frac{\text{impressed E. M. F.} - \text{counter E. M. F.}}{\text{resistance}} =$$

$$\frac{100}{10} = 10 \text{ amperes.}$$

The mechanical work done would be $400 \times 10 = $ E. $\times$ C. $=$ impressed E. M. F. $\times$ current $= 4000$ watts.

"Now suppose it is desired to run more slowly. Increase the field strength by ten per cent. The counter E. M. F. at the same speed would be 440 volts.

$$\text{The current} = \frac{500-440}{10} = 6 \text{ amperes,}$$

and the work, $440 \times 6 = 2640$ watts. But since the car requires 4000 watts to maintain the previous speed, it is now evident that it must now decrease its speed until there shall be an equality between the work required to maintain the lower speed and the work done by the motor at the lower speed due to the greater field strength. Let us learn what this speed is.

"It was seen above that work at the rate of 4000 watts

$$(= \frac{4000}{746} = 5.36 \text{ H. P.} = 176,880 \text{ foot pounds per min.})$$

must be performed in order to maintain the speed existing before the change of field strength. Suppose the car to weigh 8 tons, and suppose that on the particular track in question a horizontal effort of 25 pounds per ton is required to overcome all resistances, including those of gears or other mechanism between the armature and the axles; then a total horizontal effort of $8 \times 25 = 200$ pounds must have been exerted. This quantity multiplied by the number of feet traveled per minute must be equal to the number 176,880, representing the total foot pounds of energy utilized per minute. Hence the travel per minute

$$= \frac{176,880}{200} = 884.4 \text{ feet} = 10.05$$

miles per hour. At the new speed we must have a similar relation, viz.,

$$\frac{\text{work done in foot pounds}}{200}$$

feet traveled per minute; or since 1 watt minute $= 44.24$ foot pounds.

$$\frac{\text{work in watts}}{200 \div 44.24}$$

feet traveled per minute; hence the work in watts $= 4.52 \times$ feet traveled per minute."

It will be observed from the above that a reduction in speed results from increasing the strength of the field. The converse is also true, viz., that a weakening of the field results in increased speed. This latter can be readily proved by placing an iron bar across the pole pieces of a stationary motor, thus diverting some of the lines of force from the armature. A very noticeable increase of speed of armature will at once occur due to the weakening in this way of the field in which the armature is working.

Thus far we have assumed that the car is running upon a level track or up a uniform grade, or, in other words, that the work the motors are called upon to perform remains constant.

Of course if the load be increased, more energy is required to move it at the same speed—a greater torque will be required in the armature, and this may be accomplished by increasing the current in the latter, as already stated. The conditions for maintaining uniform speed under increased load, therefore, are obtained by a relatively large increase of armature magnetism over field strength. Thus in a shunt motor operating under constant E. M. F. the strength of the field remains constant under all conditions of work, while the armature current may vary within wide limits, and does vary directly as the load. But in a series wound motor, such as is usually employed on street cars, since all the current passing through the armature also passes around the field, the magnetism of both will continue to increase in about the same ratio until either one or the other becomes saturated, after which its magnetism cannot increase further and remains stationary, while the other increases with additional current. It has already been stated that in street railway motors it is customary to make the fields relatively stronger than the armature. This is done by making so many turns in the field magnet coils that with the currents usually employed the field magnets are nearer a state of saturation than the armature. If this be the case in a series motor, and sufficient load be thrown on to slacken the speed of the motor and thereby lower the counter E. M. F., sufficient additional current will pass to saturate the field. Should the speed become still slower, more current would pass, but the field, being already saturated, would not be further increased in strength, while the armature strength would be still further increased, thus fulfilling the conditions of relative increase of armature over field required for greater speed, and the motor would run faster until the increase of counter E. M. F. resulting from its increased speed restores a balance and the speed remains constant.

Thus far we have considered all the windings on the field magnet to constitute one integral coil. Much greater elasticity or flexibility of control may be obtained by winding the field magnet with a number of separate coils which may be connected up in various ways so as to be used separately, in series or in multiple arc, thus enabling the motorman to change the relative strengths of field and armature through a much wider range. This will be made clear by reference to Fig. 59 (see page 160) and the following explanation.

Suppose three coils of wire to be wound around the magnet cores of a motor, as shown in Fig. 59. Let the resistance of each be 1 ohm. Connect D to B and E to C (so that the three coils are in series). Suppose a difference of potential of 100 volts be maintained between the terminals A and F; then the total resistance of this field circuit would be 3 ohms. The current flowing would be

$$\frac{100}{3} = 33.3 \text{ amperes.}$$

The number of turns in each coil is two—one on each leg. There will therefore be six coils in all, and the magnetizing effect expressed in ampere turns would be $33.3 \times 6 = 199.8$.

If, however, we connect the three coils up in parallel by connecting A, B and C together and D, E and F together, the resistance of the field circuit would be one-third of an ohm instead of three ohms as before, and the current flowing would be $100 \div \frac{1}{3} = 300$ amperes instead of 33.3, and the number of turns around which this whole current would flow would be but two, and the resulting magnetizing effect expressed in ampere turns would be $300 \times 2 = 600$, or three times the former value.

If sufficient current passed through the coils when arranged in series to saturate the field, no further magnetism would be added to the field by the second arrangement, but the strength of the armature, if that were far from saturation, would be proportionately increased, as well as relatively giving the latter a greater torque as well as a tendency by reason of a relatively weaker field to greater speed; but even if the field also be far from saturation with the first arrangement, and its strength increases proportionately with the additional ampere turns of from 199.8 to 600, which would result from the second arrangement, the drop of potential required to force a given current through the field windings would be much less in the second arrangement than in the first because of the lessened resistance (in this case $3 - \frac{1}{3} = 2\frac{2}{3}$ ohms), and the electromotive force in the armature, if the latter be placed in series with the field coils, would be greater in the second arrangement than in the first, again fulfilling the conditions necessary to greater speed. It is evident that other combinations of these coils may be employed to produce different results. Thus in starting all three coils may be put in series, then one may be cut out and the other two connected in series, and next the two in multiple with each other and in series with the third, then two in multiple with the third cut out, and finally the second arrangement above described—all three in multiple. The various changes above enumerated—changes of connections—from first to last are progressively toward the attainment of higher speeds. This method of regulation is known as that by "commutated field circuits."

On most street cars there are two motors. It is evident that with two motors the commutation method of control can be still further extended by throwing the coils of the two motors not only into the above combinations with the other coils on the same motors, but by throwing the coils of one motor into various combinations with the coils on the other. Thus in starting all the coils on both motors would be thrown into series, and these in series with the armature coils of both motors, themselves in series with each other, and so by various changes gradually decreasing the resistance of the field circuits until all the field coils in each motor are in multiple with each other, and the two motors are in multiple with each other, which would be the condition for maximum speed.

CHAPTER XIX.

THE SPERRY AND JOHNSON-LUNDELL SYSTEMS.

THERE has existed quite a difference of opinion among electricians as to the relative merits of rheostat and commutated field control, but the discussions which have taken place in the American Institute of Electrical Engineers and in the technical periodicals on the merits of the two systems have apparently resulted in no conversions on either side. They have resulted, however, in making us better acquainted with the demerits of each, many of which both sides are willing to admit; but while it would seem that it makes no difference in the efficiency of a device whether the resistances be external to it, as in the rheostat control, or internal, as in the commutated field, it is impossible to get sufficient resistance into the field coils for starting purposes without sacrificing other elements of efficiency. Cars equipped on the commutated field principle are therefore liable to start with a violent jerk, because even with all the field and armature coils in series the initial rush of current is too great.

This disadvantage and the advantage possessed by the rheostat control of permitting a perfectly gradual admission of current, together with the economy of current permitted by the commutated field method, have led to a combination of the two being employed to great advantage. This combination method, known as the series-parallel method of control, is applicable where more than one motor is employed, and consists usually not in commutating separate coils on each field, but in throwing the field coils and the armature coils of the separate motors and various external resistances into various combinations with each other. It is apparent that the greater the number of motors upon the car the greater the flexibility of the system. It therefore seemed particularly applicable on the Intramural Railroad at the World's Fair, where the motor car carried four motors, and the experience on that road with the system is reported to have been entirely satisfactory from every point of view.

As before stated, it has usually been customary to equip each car with two motors. With this arrangement it is often the practice to cut one motor out of circuit entirely when the work required is small. This enables the remaining motor to be operated at about its most efficient output on loads which would be so small as to render the operation of both motors inefficient.

There are many engineers who object on theoretical grounds to the use of two motors under any conditions, on account of the tendency of any two armatures to work out of unison if there happens to be any disparity between them, either in the armatures themselves or in the strengths of the fields in which they revolve. It is very evident that if the two motors fail to work in the most perfect unison the resultant effect will be less than it should. That the perfect unison of action between two separate motors, so essential to the highest efficiency of operation, is practically unattainable is, I believe, now admitted by all, but it is approached so closely in modern construction that this method of equipment, on account of the greater facility of gearing to the axles and the greater facility of control by coupling up armature and field coils by the series-parallel method, is still almost universally employed.

There are two systems now on the market, however, which employ a single large motor flexibly geared to both axles, and the promoters of these systems claim a largely increased tractive efficiency for their methods over that possible with two independent motors. The best known of these is the Sperry system, for which the claim is made that "the motor being coupled to both axles, a tractive effort is available, which is entirely impossible with separate motors. A perfectly uniform velocity of all the wheel peripheries is obtained for adhesive effect. The great gain in drawbar pull under conditions of coupled axles as compared with the same torque applied to each axle individually is a matter not of conjecture, but of fact, the difference being more than eleven per cent. The fact is that the tendency of one wheel to slip is held back by all the others instead of by its mate only. With separately driven axles one pair may be in slipping frictional contact with the rail while the other pair is doing the work of adhesive contact." It is also claimed for the single motor system that by substituting a large single motor for two small ones the number of wearing parts and consequent loss in friction is greatly diminished, the cost of repairs, inspection and general care of apparatus materially lessened, and the commercial and electrical efficiency of the system placed far above that of any double motor equipment.

While most of these claims are theoretically true, the mechanical difficulties of gearing both axles to the same motor have been such that the theoretical advantages of the single motor have not appealed to practical men with sufficient force to cause their general adoption. On the contrary, we find to-day but one constructor building motors with this idea in view, whereas several types of single motor equipment, once advertised and pushed with considerable vigor, have disappeared entirely from public view. Vandepoele, Daft, Eickemeyer, Rae, are names that are all connected with the single motor equipment method, and there are now two more new names that must be added to this list, viz., E. H. Johnson and Robt. Lundell. These latter two gentlemen are now before the public with a single motor car equipment which seems to possess a number of features of real merit not hitherto employed. In the Johnson-Lundell system a single motor is employed, but the armature of this is wound with two independent sets of coils, each of which has its own separate commutator. There are thus practically two armatures, but since they are both revolving in the same field, they are bound to work in harmony if the coils of the two are exactly similar. This system also permits of control by the series-parallel method by the handling of the two windings as though they were separate armatures—an advantage not possessed by any other single motor system. The successive steps for increasing speeds in the Johnson-Lundell system as given to the author by the inventors themselves are as follows :

THE JOHNSON-LUNDELL SYSTEM. 173

1. Starting, external resistance, $1\frac{1}{2}$ ohms, all field coils and both armature coils in series.
2. Same as 1 with resistance cut out.
3. Field coils in series, armature coils in parallel.
4. All coils in parallel.

To give greater facility in starting, and to take up all instantaneous strains, the armature in the Johnson-Lundell system is flexibly attached to its shaft by means of springs which permit a considerable angular motion of the armature in case of suddenly applied strains before the cushioning is sufficient to cause the revolution of the shaft. . Besides permitting of a more gradual start of the car, which is effected without jar, it is claimed for this arrangement that it results in economy of current in that it permits of the generation of some counter electromotive force at the moment of starting, when of all times it is most needed and when in other arrangements it is totally absent.

Another novel feature of this equipment is the friction clutch arrangement on the armature shaft. Keyed to the latter is an iron disk and pressing against this are two other similar disks, which are forced against the keyed disc by means of a nut and spring under compression. To these latter discs are rigidly attached the two driving gears, which in this case are sprocket wheels. By tightening or loosening this nut a greater or less friction is maintained between the three disks. If the strain upon the motor exceeds this friction limit, the keyed disk will slip, allowing the armature to continue to revolve instead of stopping it suddenly, as would otherwise be the case. In practice this friction is determined by the maximum current it is deemed wise to permit the armature to take. This having been decided, the compression of the spring is adjusted by the nut so that the friction is just sufficient to permit the inner disk to slip when the dangerous current is passing through the armature.

As before indicated, the car axles are driven from the motor shaft by means of chain and sprocket gears. Whether this will prove entirely satisfactory or not in practice remains to be seen. The chain and sprocket have been repeatedly tried before on street cars, and have been abandoned as not suitable—the chief difficulty having been the rapid wear and breakage occasioned by the sudden strains to which they were necessarily subjected. When in good order, however, this method of gear is fairly satisfactory on the score of efficiency, and with the improvements introduced in the Johnson-Lundell equipment for obviating these sudden strains and shocks, and with the improvement in the chain gearing itself, the chain and sprocket may again come into favor. It is certainly the most convenient method for gearing to both axles from a single motor.

The Johnson-Lundell system calls for special attention for another reason, viz., that it is the latest attempt to do away with the overhead trolley. It is not, however, a conduit system, but rather a surface contact system, the current supply for the motor being taken from a surface conductor lying midway between the rails and flush with the street paving. This conductor is not electrically continuous, however, but broken up into lengths not exceeding eight or ten feet, which are successively thrown into electrical connection with the feeder system as the car passes over them, and disconnected again after the car passes off of them. The switching is automatically accomplished by electromagnetic devices, one of which is provided for each separate section of the trolley rail. These switches are contained in hermetically sealed boxes, usually three in a box, buried in the street beside the tracks. On board the car is placed a storage battery which is kept continually charged by the line current. The function of this battery is twofold: first, to furnish the current on starting the car necessary to actuate the electromagnetic switch by which the feeder current is diverted into the trolley rail section immediately below the car, and second, to render the car independent of the power station should occasion require, as in case it should run off the track, or where it is required to run the car over insulated portions of the track too long to be conveniently passed by the car's momentum. Since for either use but little drain is made upon the battery, the cells are not required to be of large capacity, and only a sufficient number is required so that when coupled in series their combined electromotive force will equal that of the feeder circuit. On the experimental track now in operation in New York the electromotive force of the feeder current is about 250 volts; 100 cells of the chloride battery are therefore used. These are arranged in series under the car seats, the whole being in parallel with the motor circuit. Whenever their electromotive force falls below that of the trolley circuit, a portion of current from the latter goes to recharge them. On the other hand, should the trolley circuit electromotive force drop they would come temporarily to the latter's assistance.

There is still another new system for electric railways that is destined to come into prominence in the near future, viz., that devised by Mr. H. Ward Leonard. This system is already in successful use both on electric cranes and on elevators, and in these applications has demonstrated its great economy over all other systems of control. It has been seriously objected to for street car application on account of the multiplicity of machines required, and has been unjustly condemned because it has been supposed that it would involve excessive sparking, with sudden changes of load or speed. As a matter of fact, however, the machines do not spark at all, and the efficiency of control is far greater than is possible even with the series-parallel method.

CHAPTER XX.

THE LEONARD, PERRY AND OTHER SYSTEMS.

As before outlined, in speed regulation it is necessary to vary the electromotive force as the speed, and the current as the torque or effort. In the previous methods of control this is attempted, and imperfectly accomplished by various methods of commutation of field coils and external resistances. In fact, some such method as this is all that is available where a current of constant potential, such as that now universally employed on electric railroads, is used. Mr. H. Ward Leonard, however, has taken a very bold step in his system, which consists in placing on each car a separate generator from which the car motor is directly operated—the generator itself being directly driven from the trolley current by a motor. By reference to Fig. 60 the following explanation of the system will be readily understood. We quote from a paper read by Mr. Leonard before the American Institute of Electrical Engineers, June 8, 1892:

"Each axle is driven by a gearless motor, either directly or by means of a connecting rod. The fields of these motors are excited directly from the constant E. M. F. of the line and independently of the armature circuit. Beneath the car and between the axles there is suspended a motor generator, each armature winding being in a separate field. The motor portion of the motor generator is shunt wound and connected just as a shunt motor is for use upon ordinary constant potential circuits. The field of the generator portion has its field connected across the line, and has inserted in it a regulating and reversing field rheostat. This field circuit is independent of the armature circuit. The generating armature of the motor generator is in metallic connection with the armatures of the gearless propelling motors. It will be noticed that this circuit,

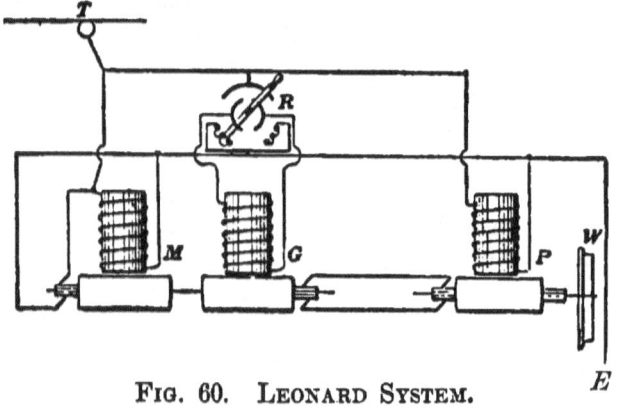

FIG. 60. LEONARD SYSTEM.

T, trolley. M, motor portion of power converter. G, generator portion of power converter. P, the propelling motor for the car. R, the regulation and reversing rheostat in field of G. E, the connection to ground. W, the car wheel.

including the armature, is a distinct and separate metallic circuit having no connection with the line in any way.

"Suppose now that our shunt motor is running at full speed, and that our controlling rheostat in the generator field circuit is at its central position, so that the generator field circuit is broken. Although the generator armature is being driven at full speed, it is revolving in a field having no magnetism except the residual magnetism, and

THE LEONARD SYSTEM.

hence produces practically no volts. Let us now move our controlling switch so as to place the generator field across the line, but with a resistance in series with the field of ten times the resistance of the field coils. We now give a slight excitation of the field and a development of volts at the brushes of perhaps forty volts. This voltage will produce a current through the armatures of the driving motors dependent upon the ohmic resistance of this circuit only; and hence, even at this low voltage, a large current will be produced, which, being in a field of full strength, will cause a torque sufficient to start the armature. The speed of the armature will, of course, be governed by the counter E. M. F. which its revolution produces in its strong field; and hence, just as in the case of a shunt wound motor, its speed will be practically constant so long as the E. M. F. supplied is constant.

"If we now gradually increase the magnetic field of the generator by cutting out resistance by moving the controlling switch, we will gradually raise the E. M. F. of the armature circuit, and with it the speed of the driving motors. Since these armatures are revolving in a constant field, the torque they produce will be exactly proportional to the current in them, and the current will automatically flow exactly as is required to produce the necessary torque to maintain a speed such that the counter E. M. F. will approximately equal the E. M. F. supplied by the power converter. Thus it will be seen that the speed of the car will be dependent upon, and proportional to, the E. M. F. supplied by the power converter, and the torque or tractive effort will be dependent upon, and proportional to, the current supplied by the power converter."

While this method has not as yet been actually

introduced on electric street railroads, its feasibility and economy and the facility of control afforded have been amply demonstrated on numerous electric cranes, elevators and hoists, and in connection with machinery of various kinds requiring the greatest nicety of speed control under widely varying loads. In such applications economy of space has not been the prime requisite that it is in street car traction, and the fact that the arrangement seems necessarily bulky, with the three machines involved, and the erroneous idea which has been general that the sparking would be excessive under working conditions, has militated greatly against its introduction in street railway work. As before stated, those who have examined into the workings of the system, among whom may be included the author, aver that there is practically no sparking even under the most favorable conditions for the same, and the author is advised that changes in detail (though not in principle) have already been partially perfected which will obviate the real difficulty that now exists, viz., bulkiness.

Mr. Leonard has also adapted this same system to the operation of cars from alternating current circuits. In this application a synchronous motor generator is substituted for the direct current motor generator employed in the direct current method. Those who wish to read a full description of this system are referred to vol. xi. of the Transactions of the American Institute of Electrical Engineers under the title "How shall we Operate an Electric Railway extending One Hundred Miles from the Power Station?" by H. Ward Leonard.

THE PERRY SYSTEM OF SERIES ELECTRIC TRACTION.

Thus far the method of distributing current for electric traction has been that by constant potential. That is to say, the pressure at the dynamo terminals has remained constant, and the current in amperes delivered to the moving cars has varied as the requirements. This method as a whole has thus far proved the most flexible, but with the increasing distances to which our electric roads are constantly reaching it is becoming less and less satisfactory. We have already adverted to the drop in potential at the further end of the line where the current is used by the present methods. This drop can be obviated in either of two ways, either by placing sufficient copper in the feeders to reduce it to a bearable amount, or by increasing the potential. If the potential be raised sufficiently to operate a car properly at the further end of a long line, it will be too high for the same car when nearer the power station, so recourse is had to additional copper. But the amount of copper that can be thus used with an initial pressure of five hundred volts is soon limited by commercial considerations, beyond which it becomes more expensive than to erect and operate another power station at a distant point. Just how far it is profitable to operate a road from a single station must be decided for each particular case, but it is probably within the truth to say that where the distance to be reached exceeds six or seven miles it will be cheaper to build a second power house than to put sufficient copper in the feeders to render those greater distances practicable. In the constant potential method of distribution the varying demands for energy are met by a varying quantity of current, yet our conductors, which are fixed in size, are properly proportioned only for a given current. For that particular current they are of exactly the proper size; but for any other current, be it larger or smaller, the conductor used is not the most economical; it will contain either more copper or less than that which can most economically carry it. If, however, we should use a constant current, and vary the energy transmitted by varying the electromotive force, we could proportion our copper to that current once for all, and since by the conditions imposed the current does not vary in quantity, the size of our conductors need not vary whatever the amount of energy they are required to carry or whatever the distances to be reached. Illustrations of this method of distribution are seen in our arc light circuits. Many of these already extend upward of 20 miles and carry from 60 to 100 or more 2000 C. P. lamps, yet the conductors on such circuits are no heavier than would be necessary for a circuit 1 mile or 5 miles in length carrying but a single lamp or 4 or 5. To render the same sized conductor equally economical for the transmission of large amounts of energy to long distances, it is only necessary to raise the electromotive force correspondingly. Large amounts of energy can thus be much more economically transmitted to long distances by the constant current method than by the constant potential method, the economy becoming particularly conspicuous where the demands are variable and where the distances are also variable, as is pre-eminently the case in electric traction.

On account of these and other advantages of the constant current method of distribution many attempts have been made to adapt it to street railway purposes. But there have been difficulties in the way of its adaptation to this use which until they were removed introduced greater objections than were the benefits sought to introduce. The writer, however, has invented a system which he believes obviates all the difficulties (see Trans. Am. Inst. Elect. Engineers for 1892) hitherto thought to be inherent in the constant current method, and which at the same time sacrifices none of the advantages sought to be gained. His improvement over previous methods consists simply in supplying to the railway circuit the same device by which the series arc light system was converted from a failure to a commercial success, viz., an automatic cut-out by which on the failure of an operating device it is automatically cut out of circuit. Without this invention the practicability of even two or three arc lamps in series was uncertain ; with it any number may be practically operated.

Professor S. H. Short experimented quite extensively a few years ago with the constant current method of distribution for street cars, and had for a time two roads in more or less successful operation—one, about three miles in length, between Huntington, W. Va., and Guyandotte, and the other, of about the same length, in the city of St. Louis. He had not conceived the idea of the automatic cut-out, and found it impracticable to operate more than three cars simultaneously on his lines. He, however, demonstrated the economy and efficiency of the system up to its practical limit. With the addition of the cut-outs and the other radical changes introduced by the writer it is believed that the new system possesses all the

flexibility of the multiple arc method with the additional advantages of the series method, and in recognition of these claims the Franklin Institute of Philadelphia recently awarded him the John Scott legacy premium and medal.

This system may be briefly described as a double trolley system, divided up into sections of greater or less length, according to the headway of the cars provided for. Each of these sections is fed independently from a common feeder wire by means of electromagnetic devices inclosed in hermetically sealed boxes along the side of the track very similar in design to those employed in the Johnson-Lundell system described above. They are not nearly so numerous, however, as in the suburban districts, where the distances between cars is great, not more than one or two to the mile are required. In the cities, however, where the headway of cars is much less, their number must be increased proportionately.

But one car can operate on a section. If a car runs onto a section already occupied, both cars will become inoperative until one of the two pulls down its trolley—thus constituting this a perfect automatic block system. All cars on the system are in series with each other, and since the current strength by which they are operated is invariable, their speed may be checked on descending a grade or in bringing them to rest by reversing the connections. This reversal of the brushes converts the motor into a dynamo, which, being in series with the one at the power house, contributes energy to the line to the last turn of the wheels. In this way the energy absorbed in ascending grades or in starting from rest is thrown back on the line for use elsewhere, instead of being frittered away in heat on the brake shoe, as in present methods. The method of control of cars operated by this system is ideally simple. There are neither commutated fields nor external resistances to be considered, so that with this system the advocates of both methods of control join hands, since the method involves the disadvantages of neither. In the series system the car is started, speeded up, slowed down and reversed by simply shifting the position of the brushes on the commutator. The equipment of the generating or power station is also extremely simple, requiring for each unit nothing more than a voltmeter, an ammeter and a line switch by which the circuit can be closed or opened, no rheostats, bus bars or complicated switchboard systems being required. This system is thought to be peculiarly applicable to long lines, such as suburban and interurban lines, and it is believed that the time is not far distant when its advantages on such lines will be fully appreciated.

STORAGE BATTERY TRACTION.

A self-contained car or one that is self-propelling would be an ideal were there not attendant disadvantages of such a serious nature as to overshadow the advantages possessed. The storage battery seemed to hold out such bright prospects of success that many earnest efforts have been made almost from the first invention of the storage battery by Planté and its improvement by Faure down to the present day. It, however, has been a disappointment in every case, in this country at least, where it has been tried, notwithstanding the fact that neither engineering skill of the highest character nor expense has been spared to make it a success. The causes of these failures are many and seem to be inherent in the battery itself as at present constructed. We know of no more efficient method of storing electricity * than by means of lead plates. These are so heavy that a street car equipped with sufficient battery capacity for its successful propulsion must carry in this form alone more weight than it is possible to put upon it in the shape of passengers. The empty car is therefore handicapped on starting out with a non-paying load greater than that from which it can expect to derive revenue. This means not only additional expense for haulage, but is destructive to track and cars alike. Since at least two sets of battery equipments must be provided for each car, which will greatly exceed the cost of motor equipment, the fixed capital investment upon which interest must be earned is excessive. Another disadvantage of the battery is that not more than about seventy per cent. of the energy put into it can be drawn out of it again for use, and even this amount is available only when the batteries are new and in good condition. When they are old, the percentage of energy used for charging that is available for use is very much less than this, running down to fifty, forty, thirty per cent., and even less. The deterioration of the plates, too, is very rapid in street car work owing to the jarring of the cars and the swash of the liquid, causing a loosening of the active material, its falling out and causing short circuits, resulting

---

* It is not *electrical* energy that is stored in the storage battery, as is popularly assumed, but *chemical* energy. The term "electrical storage" has come into such general use, however, and it is so convenient, that with this explanation I may use it without creating confusion.

in the destruction of the plates. The wear and tear or depreciation account is, therefore, excessive, and, taken altogether, experience has almost conclusively pointed to the inadaptability of the storage battery to traction purposes.

The storage battery, however, permits of the most economical speed control of any known, as by commutating the cells into groups in series and in parallel, by using some for separately exciting the fields, while various other combinations are used for feeding the armature, an ideally economical speed regulation is obtained, somewhat similar to that advocated by Mr. Leonard, but without the multiplicity of apparatus or external resistances used by him.

Should a lighter and more durable storage battery than the present type ever be invented it is not unlikely that it would come into general use for traction purposes, but with our present types it seems exceedingly unlikely that it can make much progress in this direction.

CONDUIT SYSTEMS.

The unsightliness of the overhead trolley and its obtrusiveness in the streets are its chief objectionable features. To overcome these many attempts have been made to carry the wire underground in a conduit similar to that used in cable railways. It would seem, at first blush, a very simple thing to place the trolley wire in such a conduit and have it work at least as well as it does overhead, for that is all that is asked of any conduit system, but difficulties have arisen that have militated against the success of the conduit system thus far that have brought it into disfavor.

The chief difficulty, to my own mind, is the greater expense involved in such a plant. It is but natural that parties having railway franchises should desire to avail themselves of them by the least expensive and most efficient means where those two qualities are not incompatible. It so happens that the overhead trolley does possess both of these qualities, hence since there is no advantage to the investor in putting the trolley wire below the surface of the street, but, on the contrary, a disadvantage in the way of greater expense, capital has not had the incentive to investment in conduit systems that have been offered by the overhead trolley. In our larger cities, however, the public is already clamoring for the burial of the wires, and with this incentive many inventors are at work endeavoring to devise a practicable conduit system which shall not at the same time be too expensive either in installation or in operation.

The chief difficulties to be overcome in conduit work are to maintain an efficient insulation of the trolley wires from the ground or conduit itself; to prevent the insulation of the trolley wire from the trolley contact by dirt entering through the slot; from mechanical and electrical difficulties, switches, etc.; and to provide an efficient contact between trolley and wire under conditions which prevent the entrance of dirt into the slot.

The two best known examples of successful conduit construction are the Siemens and Halske and the Love systems, in both of which the principle of the overhead trolley is followed closely. The conductors, however, are not placed directly beneath the slot, but off to one side, where they are protected by the roof of the conduit from in-falling dirt, water or snow. The traveling contacts are made of proper shape to reach around to the conductors, with which they are brought into contact when desired. In the Siemens and Halske system, which has now been in successful operation for several years abroad, the conduit and slot are at the side of the track. In the Love system, which has been tried in Chicago and is now in successful operation in Washington, the conduit is in the center of the track. While these two systems differ considerably in details, they are the same in principle, which is exactly that of the overhead trolley, special attention being given to the insulation of the trolley wire from its surroundings, to the adequate drainage of the conduit itself and to adapting the trolley to its new requirements.

Other inventors have endeavored to solve the insulation problem somewhat on the same lines as those adopted in the surface contact system of Johnson and Lundell, namely, by dividing the trolley wire up into short sections which are normally dead, but which are successively thrown into electrical connection with the live feeder wires as the trolley passes onto them, and thrown out of connection with them as the trolley passes off from them. By thus having only that portion of the trolley wire actually in use active, the tendency to leakage of current is of course greatly lessened, but there is introduced in its stead a multiplicity of switches upon whose proper working the success of the system depends, which is of course a source of weakness, or possible weakness, scarcely less desirable of avoidance than leakage itself. Great ingenuity has been displayed in designing systems on this plan, but none of them has been in practical use long enough to demonstrate beyond question its absolute practicability.

Still another type of conduit known as the closed conduit has been devised in which by various means, usually by electromagnetic devices, the passing car maintains a supply of current by attracting to short surface conductors the live conductors buried beneath the surface of the street. Sometimes this contact is made by the pressure of the car wheels or their flanges upon spring contacts over which they pass, but in all cases the methods of speed control are identical with those employed on the overhead trolley systems.

CHAPTER XXI.

THE MANAGEMENT OF STREET RAILWAY MOTORS.

At the outset of this chapter the author might as well confess his inability to teach a novice, either by letterpress or pen, how to actually manage an electric street car in its various moods and whims, if so they may be called. If he can impress upon his readers at the outset that motors do not have whims at all, and that the seeming irregularities of their behavior are all due to definite causes which are in most cases within the control of the motorman, he will have made a good start. No one from book reading alone can expect to step full-fledged a motorman upon his car. There is much, after all is told, that must be learned by actual experience. For instance, the *sound* of the motors often tells one much, and not only indicates either that everything is all right and permits him to continue on with confidence, or that something is wrong, and, moreover, frequently indicates just *what* is wrong, or so nearly so that the experienced motorman need not look far to find the trouble, whereas the novice might spend an hour or two fruitlessly hunting for the fault. And this was the moral intended to be taught by the little anecdote at the beginning of this book.

We believe, however, that if one thoroughly understands the construction of his apparatus and the principles upon which it operates, the instructions which follow will well supplement a growing experience, and enable the conscientious operator to avoid many difficulties, and to correct them, when they do occur, the more readily. It was with the object of *leading up* to the management of the motor rather than of teaching it that this work has been undertaken. It only remains for us to give a little kindly advice, which it is hoped the reader will by this time have been prepared to understand. The remaining pages will therefore partake more of the nature of those medical works intended for family use than of the nature of a technical medical treatise.

The first thing to keep in mind is to keep your motor dry, for if it get wet the whole structure is liable to give way—burn out. The second word of advice is to keep it clean, for thereby most of its ills will be avoided. Water and dirt are the two greatest enemies of the electric motor. The third is, study the wiring of your car, so that you have clearly mapped out in your mind the various connections and their functions. This will enable you to test out a fault which could not easily be detected by the eye, and to locate it, at any rate, within a given circuit. In giving this last advice to the motorman the author is not unmindful of the difficulty, nay, even the impossibility, of the motorman's being able to trace out the wiring of his car unassisted. To do this he must have the full co-operation of those in authority; and here we have a word of advice to give to those in charge of the rolling stock.

If a motorman or other employee of yours, having in the performance of his duty to do with your motors, wishes any information in regard to the same which it is in your power to supply,

supply it fully and freely. In fact, it is your duty, if you would have good service, to educate your employee to the highest degree possible in his duties. Do not be content to let him do things with your machinery simply because you tell him to, thereby making of him a machine which is even more likely to get out of order than the inanimate machinery he has to handle ; but after telling him what to do and what not to do try to explain to him the reasons therefor, and the penalty —not to him, but to the machinery in his charge— of disobedience. Encourage him to ask proper questions—and all questions in regard to his motors are proper ones—and help him to become an intelligent man ; for by so doing alone can you get the best service, the most for your money.

The next advice to the motorman is that he study the " habit " of his motors. Let him keep his ears open to every sound until any variation from the same would awaken him even if he were asleep, for the sounds given out by the motor are as surely an indication of its condition as is the pulse of a human being of his state of health. A variation from the normal sound is often the first indication the operator may have of trouble with his machines. If heeded at once, disaster may be entirely averted where it might otherwise almost surely follow.

If the ear detects anything unusual, the car should be stopped at once and a careful examination made to detect if possible the cause. If it cannot be located at once, it may be well to cut out first one motor and then the other, running the car carefully for a short distance with each separately, if the grades are such as to make this safe. In this way the trouble may be located by sound in the motor in which it exists, and thus its specific nature and exact location be more easily traced.

The most common diseases of electric motors of any kind, street car motors included, their symptoms and remedies, are the following:

First of all comes

SPARKING AT THE COMMUTATOR.

A properly constructed motor in normal working condition should not spark at all, or at least not noticeably. Sparking may therefore be regarded more as a symptom of a disease than as a disease. The sparking of the commutator is often the first indication that the operator has that everything is not as it should be. When sparking is observed, therefore, an investigation should be made at once to determine its cause and to rectify it on the spot, if possible, or, in case it is not possible to do this, to run the car into the shop for repairs. Sparking should be stopped for its own sake, however, for if allowed to continue it will corrode the commutator blocks, and in this way increase itself until the commutator is so far gone as to require renewal.

Crocker and Wheeler, in their most excellent little work on "The Practical Management of Dynamos and Motors," assign fourteen different causes for sparking, not all of which, however, need concern us here. The following, however, are those which especially concern the motorman:

First Cause.—Armature carrying too much current. This means that the motor is being overworked. The motorman need not be alarmed if his motor sparks some on ascending a heavy grade, for that is rather to be expected, especially if the load be at the same time heavy. The remedy for this, of course, is to save your motor as much as possible by gaining momentum before reaching a grade, and then maintaining a uniform slow speed until the grade is passed. A motor that sparks on ascending a grade should never be stopped on the grade, if it can by any possibility be avoided, as on starting again the work it will be called upon to do will be many times that which it ought to do, and disastrous results may follow.

But the sparking from overload may not always be due to the excessive *useful* work the motor is performing. It may be due to the striking of the armature against the pole pieces, to the binding of the armature shaft in its bearings, to a bad short circuit or to the grounding of the motor on the frame. Any of these latter causes, if active, are likely to cause sparking when the motor is not doing much apparent work, and this fact will help to distinguish which of the two classes of troubles causes the overloading. The general indication that the sparking is due to overload is the overheating of the whole armature. If this overload is due to frictional causes, they may be detected by examination first of the bearings, which will be unusually hot if the trouble lies there, and second by an examination of the armature. If friction is indicated there, the trouble is extremely serious, as in overwound armatures (viz., those on which the coils are wound on the surface as distinguished from the iron-clad armatures, in which the coils are placed in slots on the surface, and therefore beneath the surface) continued friction is sure to wear off the insulation and cause a burn-out of the armature. In this case the motorman should exercise his mechanical ingenuity, and so center his armature that it will not strike the fields at all.

Second Cause.—Brushes not set at the neutral point. We have seen that there are two positions in every revolution of a coil in which it generates no electromotive force. In a two-pole machine these two positions are diametrically opposite each other, and in a four-pole machine they are 90° from each other. These points are called the neutral points. If the brushes bear at exactly these points, there should be no sparking, but if they bear at any other points on the commutator the brushes will pass off from bars that have an electromotive force which is greater the further the brushes are removed from the neutral points, and it is this electromotive force that causes the sparking. We have seen that as the current supplied to the motor changes, so do the positions of these neutral points in some machines, so that it is necessary in such to move the brushes back and forth as the load varies. It is this change of the positions of the neutral points that causes the sparking due to overload, just described. But the ampere turns on street railway motors on the armatures and fields are so disposed, the one predominating over the other to such an extent that the load may vary within wide limits without perceptible change of the neutral points. The brushes are therefore usually fixed once for all at the proper places, so that the sparking, if due to the brushes, is probably due to one or more of the following causes rather than to wrong position:

a. Commutator rough, eccentric or has one or more high bars, or what are termed *flats.* To detect these the commutator should be examined while at rest for roughness and also for eccentricity. This latter can be detected better, however, by watching the motor carefully when slowly in motion. If the brushes alternately rise and fall, the commutator is not centered properly on the axle. When running fast, the whole armature may *chatter*. High bars or flats—the latter being flat surfaces on the commutator—are best detected while the motor is running by resting the finger nail against the commutator. Any irregularity of surface will thus be readily detected. These are all difficulties with which the motorman would better not fool, for fear of increasing the trouble. His duty will have been done if he cuts out this motor and proceeds to the car barns at once with the other motor.

b. Brushes make poor contact with the commutator. Close examination will show that the brushes touch only, at one corner or only in front or behind, or there is dirt on the surface of contact. The remedy for this trouble readily suggests itself. Clean the commutator and replace the brushes, being careful that they have ample bearing on the commutator. Occasionally the fault lies in the brush itself: it may be extremely hard, or have extremely hard spots in it which wear away less readily than the remainder of the carbon. The remedy for this is to throw away such brushes and replace them by new ones.

Sometimes a good brush has worn unevenly through grit on the commutator, and only needs dressing down to give good service. This is best done by drawing a strip of sandpaper back and forth between it and the commutator while the brushes are pressed down. This will dress their bearing surfaces to fit the commutator properly.

Third Cause.—Short-circuited coil in armature.

This may be caused by a little carbon dust or other conducting material getting in between two of the commutator bars or between the connections leading to the bars. Perhaps the best indication of a short-circuited coil is the increase of heat in that particular coil or coils on the armature. If in feeling around the surface of the armature one or more coils appear much hotter than the rest, a short-circuited coil is to be suspected, and a careful examination of the commutator and its connections should be made to discover the cause, and if found it should, of course, be removed. If the cause be removed early in the trouble it may have done no harm, but a short circuit is very likely to cause a burn out of the armature. If the trouble is in the commutator or its connections, its remedy, if taken in time, is exceedingly simple, but the short circuit may be in the armature itself, and if so cannot be corrected by the motorman. If he have reason to suspect that it exists in the armature, his only recourse is to cut that motor out and proceed to the stables with the other motor. The short-circuited coils will there have to be replaced by new ones.

The same effect may be produced exactly by a "ground" on the armature, which together with the intentional ground forms a short circuit. The indication, aside from the unduly heated coil, is very bad sparking occurring at intervals.

Fourth Cause.—Broken circuit in armature. This is usually indicated by violent flashes like the preceding, but unaccompanied by the heating of the coil, the flashing, as before, occurring at intervals when the commutator segment belonging to the broken coil passes under the brushes. The flashing in this case will be very much worse than in the preceding case, even when the motor is running slowly. Examination should be made to see that the flash is not due to a high bar or dirt or other insulating material on one of the bars. If not due to either of these, the break is most likely to be found in the connections between the armature coils and the commutator bars. If it be due to a broken commutator connection, a temporary remedy is found in connecting the disconnected bar with its neighbors by driving in between the bars a piece of copper wire so as to short-circuit the broken coil. If the break be in the coil itself, rewinding is probably necessary, and the motor should be cut out of circuit at once.

Fifth Cause.—Chatter of brushes. The commutator sometimes becomes sticky when carbon brushes are used, causing friction, which throws the brushes into rapid vibration. When this is the case, it is readily detected by the tingling or jarring sensation produced on the hand when lightly placed on the brushes. At the first opportunity the commutator should be cleaned with a rag or waste and oiled slightly. This will stop the trouble at once.

Sixth Cause.—Flashing all around the commutator. This may be due either to particles of carbon between the bars or to broken coils, or both, and the remedy is that recommended before for such troubles. If, after cleaning the commutator and no breaks are found, the flashing continues, the motor should be disconnected and the car run into the shops for overhauling.

MOTOR STOPS OR FAILS TO START.

First Cause.—Great overload.
Second Cause.—Very excessive friction due to

shaft, bearings or other parts being jammed, or armature touching pole pieces. In either of these cases the armature would most certainly burn up were it not for the fuses, which are intended to melt and break the circuit before sufficient heat can be generated in the armature coils to do damage. A careful examination should be made to see what the trouble is, and it should be rectified, if possible, at once.

Third Cause.—Circuit open, due to : (*a*) Safety fuse melted, (*b*) connection to motor broken or slipped out of binding post, (*c*) brushes not in contact with commutator, (*d*) hood or canopy switch open, (*e*) broken or imperfect contact in controlling rheostat, (*f*) failure at generating station. Trouble due to any of these causes is indicated if the car fails to move when load is removed or when load is light. In such cases the current should be turned off immediately at the hood switch, and the break looked for as indicated. The lamp circuit should be turned on. If lamps burn or other cars are found to be moving, the trouble does not lie with the generating station.

CHAPTER XXII.

SPECIFIC DIRECTIONS TO MOTORMEN.

SOME of the electrical companies furnish, and all of them should furnish, a list of empirical rules for the management of their motors. It is too often that the motormen never see these rules at all, but receive them by word of mouth from the foremen or whomever they look to for instructions. It is very desirable that each motorman should have these rules in convenient form for reference, and they are herewith reproduced. They are essentially the same for all makes of motors, and are as follows :

1. In taking a car out of the barn where it has been standing with trolley off put on the trolley, place handles on controlling stand and see that the current is off ; then throw in the hood switch.

2. On most modern equipments there are two levers—one for reversing, the other for controlling the current and speed of car. In starting move the controller quickly from right to left until you feel the contact touch the first point, and then slowly move it farther until the car moves. Allow the car to gain a little headway, and then move on as the car gains headway till the lever is as far as it will go. Usually the controller switch should be allowed to rest on each successive notch long enough for the car to gain the headway due to that combination before it is moved to the next

notch. In the Westinghouse control the first notch throws the whole resistance and the two motors in series. The second notch cuts out half the resistance, and the third notch the whole resistance. The fourth notch throws both motors in parallel with one another and each in series with half the resistance. The fifth notch cuts out the resistance of one motor, and the sixth or last notch cuts out the resistance of the other motor, leaving them in parallel, which gives the greatest speed and highest efficiency. In this arrangement, which is a series-parallel arrangement, the first two notches are merely starting points, and the handle should only be allowed to rest momentarily on each. The third is a good slow speed running notch. To obtain greater speed or more pressure the handle should be moved slowly, but continuously, from the third to the fourth notch. This and the next notch should be used only momentarily, not steadily. In this, as with all other arrangements, the last notch should be used for heavy work or fast running.

3. Before leaving the car barns, however, the motorman should examine carefully the grease cups and see that they are filled. Examine brushes and motors, making sure they are in fit condition to begin the day's work.

4. With hood switches open, try the reversing lever and controller lever to make sure they are in good working order. Then, with controller lever off, close hood switches; note that trolley is on the line and that you have a current by lamps being lighted. Now move controlling lever around slowly. If car moves off, all is probably right. If car refuses to move after controlling lever is moved to last notch, turn lever back to "off"

point, get down from car and note that rail is clean, that the fuse plug is in place, that the ground wire from motors is attached to truck frame and that the cut-out switches for both motors are closed. If the above conditions are fulfilled, then examine car wiring for a broken wire or a loose connection. Failing to find the trouble, report to the car starter or person in charge.

5. If the General Electric Company's series-parallel controller, form K, is used, the motorman's attention is directed chiefly to noting that everything about the switches and contact points and cable connections in the controller are in good order. Two switches are located in the lower part of these controllers. The one to the right when thrown up as far as it will go cuts out motor No. 1, or the motor nearest the fuse box; the switch to the left when thrown up cuts out motor No. 2, or the one farthest from the fuse box. A small quantity—enough only to form a very thin film—of vaseline should be used on the contact strip in all controllers to prevent any cutting or wearing.

In both controllers of this type the upper cut-cut plug cuts out the same motor, which is designated as No 1. The motorman should find out which is motor No. 1, so that he will be able to remove the proper cut-out in case of trouble without experimenting. It is recommended that the number of each motor be painted on it where the motorman can see it. These cut-outs are designed to be used only in case of trouble. When either plug is removed, the starting of the car is delayed until after the controlling handle has passed the third notch. *Hence, in starting with one cut-out plug removed, throw the handle directly to the fourth notch.*

6. The reversing switch determines the direction in which the current shall flow through the motor fields when it is turned on by the controlling handle. In the Westinghouse apparatus, for instance, the reversing switch has three notches. The central one, at which the handle is placed, cuts off all current from the motor fields, so that in this position operating the controlling handle has no effect. When it is desired to start the car, first see that the controlling and reversing handles are at the "off" position; second, close canopy or hood switch; third, throw the reversing switch forward or backward, according as it is desired to go in one direction or the other; fourth, throw the controlling handle, and the car will start.

Throwing the reversing switch entirely over reverses the direction of the current in the fields, but this should never be done unless the controlling handle is at "off," otherwise the rush of current through the coils which will be due to the *counter electromotive* force of the motor *added* to that of the line and that due to the discharge of the magnetism of the fields will be so great as to endanger the coils.

7. When throwing the controller arm to "off," the movement should be rapid, especially in passing the first point, so as to avoid drawing an arc.

8. If the controlling arm should go hard or stick, do not force it, as this would only make matters worse. Pull down the trolley or open the canopy switch, or, better, do both, then remove cover of controller. An inspection will probably show that the trouble is due to want of oil, roughness of contacts, or something of this kind, which can easily be corrected.

9. Never run with trolley in wrong direction

except in cases of extreme necessity, and then very slowly.

10. Never stop car so that the trolley wheel will be directly under a circuit breaker in the line.

11. Always have current shut off when trolley wheel is passing over a circuit breaker in the line, else the wheel in passing off will draw an arc, which tends to damage both line and wheel.

12. Never leave car without removing the controller handles and opening canopy switches.

13. Never reverse the motors when the car is running except in cases of extreme necessity, such as avoiding a collision or to save a life. In these cases reverse the current in the controller, keeping the handle on the first or second notch until the car begins to move backward. Remember that if reversal takes place with controller at too high a notch the wheels will lose their adhesion to the rails and spin around backward, and the car will not stop so quickly as if they kept revolving in a forward direction.

14. Go around curves slowly, using third notch.

15. When entering a turn-out or curve, the conductor should be on the rear platform and should have the trolley rope in hand.

16. Slow up at all street and railway crossings, at all rough places in the track, and pass overhead switches with the current thrown off.

17. It is better not to stop on very heavy grades, or on or just before entering curves, if it can be avoided, on account of the extra current required for starting up again under such conditions.

18. Ordinarily in stopping the car always release the brake somewhat, just before the car comes to a dead stop. Do not let the brake fly, or kick the brake dog off, for if you do the armature will take

up the lost motion in the gears, and when starting again it will necessarily be with a jerk. This is unpleasant to the passengers and hard on both motors and gears.

19. Do not keep brakes on in rounding curves. This has been advocated, but is wrong and involves a useless waste of power at the worst possible time. It is one of the commonest and worst mistakes motormen make. It is well to have the brake in hand so that it can be instantly applied if necessary, but it should be entirely " off."

20. Motormen should never run the car when the trolley is off, especially down grade, for if the brake should fail he could not reverse.

21. In descending a grade it is best to run slowly, for should it be necessary to stop suddenly it would be impossible to do so if the speed were high.

22. If, in wet weather, when climbing a grade the wheels slip, gradually work the controller arm toward the first point, throwing it to the position of " off " if necessary, until the wheels get a grip, then work the arm gradually over toward " full power " again.

23. In applying brakes on down grade be sure not to allow the wheels to get to slipping, for when they once commence to slip or " skid " they are of very little use in stopping the car. Many accidents have occurred in this way. This precaution is especially necessary where stops are made on a descending grade. Should the wheels begin slipping, however, better let the car run faster for a few moments until they get hold again, and then apply the brakes gradually until the car is under control.

24. Run slowly through flooded places, if possi-

ble, with current cut off. When examining motors, never allow water to drip from clothing or hat into the motors.

25. If car won't start on dry or dirty rail, put controller arm on first or second point and rock the car. If this fails to accomplish the purpose, have conductor take a piece of wire or switch stick and rub one end of it against the rear tread of the wheel while the other end is pressed against the rail. In case an uninsulated wire is used, *break contact at the wheel first,* keeping the other end against the track, else a shock will be received.

26. Never attempt to put in a new fuse unless canopy switch is open or the trolley is off; otherwise you may get a shock and damage the fuse connections also.

27. Should it be found impossible at any time to start the car, try the following until the trouble is located:

a. If there is no evidence of current, notice other cars. If they are all right, the trouble is in your own car.

b. Throw on lamp circuit. If the lamps light up, the trolley and ground wires are all right. Now work controller, and if the lights go down or out the trouble is probably due to poor contact between the wheels and the rails (try 25), or the section of track on which the car is standing may be "dead." Use a longer wire and connect wheel with another rail, as in 25.

If lamps do not light, examine lamp fuse box to see that fuse is not blown, and make sure that ground connection is not broken. Make sure that lamps have good connection in the socket. If they still fail to light, you may be reasonably sure the power is off.

208   ELECTRIC RAILWAY MOTORS.

c. Ascertain whether the fuse has been blown. If so, throw canopy switch and put in new fuse (26).

d. See that both motor cut-outs are in place.

e. Try both controllers, and if one works the trouble is probably due to poor contact in the other. In this case throw canopy switch, remove the cover of the controller and examine the contact blocks to see that they all make proper contact.

f. Examine the brushes of the motors to see that they are not broken, and that they make good contact.

28. In case current is shut off at station for any reason while the car is running, bring controller to "off" position immediately. Then turn on light circuit, and wait until the lamps light up; when they have reached their usual brilliancy, *but not before,* start the car. The reason for this precaution is that, should you turn the controller far enough to start the car before the full current was on, there would be little or no counter electromotive force generated to keep back the rush of current when it did come, and your armature might be injured either by heat or by the sudden jerk that would result.

29. *In case the brakes of a car fail to operate* there are two methods of stopping the car by the use of the motors. The first consists in reversing the direction of the current in the motor fields as follows:

a. See that the controlling handle is at "off."

b. Reverse the reversing switch.

c. Throw the controlling handle around to the first or second notch — never beyond the third notch, unless the fuse blows. In that case (Westinghouse) move the handle around to the last notch and leave it there. This converts the motor into a generator and it will come to a stop if on level, or if on a grade will slacken up. In other makes (Short) the simple pulling over of the reversing lever after the controlling lever is turned to " off " will accomplish the same thing.

The second method will operate successfully whether the trolley is off or on the wire. It is as follows :

a. Place controlling handle at " off."
b. Throw canopy switch to " off."
c. Reverse the reversing switch.
d. (Westinghouse.) Throw controller handle around to last notch and *leave it there* until the car stops. The step c converts Short motors into generators. The additional step, d, is required in Westinghouse motors.

When cars are running away downhill, the method of short-circuiting the motors on themselves and thereby converting them into generators is recommended as a last resort.

30. In case a motor bucks or flashes, examine brush holders, and if they are covered with dirt or mud, open canopy switch and clean them. A loose joint in the circuit between the brushes and the field will almost invariably produce flashing ; trace circuit carefully and find it. See that the spring is tight and that there is no dirt coating on the bearing surface of the brush.

31. If the trouble is not found with the brush holder, and motor has a peculiar smell like burned rubber or shellac, wait for the next car to push her in.

32. Motormen should always have a wrench on car to tighten up a loose nut, and should be

constantly on the lookout for troubles of this kind.

33. Always report to inspector any trouble with track, such as "dead" rail, or of trolley wire, such as break of line or insulator, etc., and any unusual noises of motors, having first, however, endeavored to account for these latter yourself.

34. When you desire to run at slower speed and controller is full on, it is usually considered better to first pull it clear back to starting point and then back to position you think will give the desired speed.

35. Never pull reversing lever over while controller is on. If you do, you are likely to blow the fuse or burn up the motor, in either case losing control of your car.

36. If in running along you feel the car suddenly let up, throw controller off and ascertain the cause if you can. It may be the trolley is off, passing a trolley break, a fuse blown or current cut off at the power house.

37. If you do not find any trouble, try to start again. If the car does not move, proceed as directed under such circumstances.

38. Do not run over sticks, wire or other obstructions on the track, as they are liable to get entangled in the motor. Get down and remove them.

39. If paving blocks or other projections stick out above the pavement, slow up and be sure the motor will pass over without touching before attempting to pass. Of course you will remove the obstruction, if possible.

40. In case of the repeated blowing of the fuse without apparent cause, pull down your trolley and wait to be pushed in.

41. The proper handling of a car on a curve is perhaps the most difficult task that the new motorman has to learn. A good rule is the following: In approaching a curve cut off your controller, and bring the car down to a slow walk before entering, and have your brake in hand, but free, unless it be down grade. This will let the car run into the curve easily and without shock. As soon as you feel that the car is fairly on the curve apply sufficient current to carry the car around the curve at about the same rate of speed, cutting it off again just before leaving the curve. This will allow the car to take the tangent with the least possible shock.

Always bear in mind that *anything that causes the car to jerk* is wrong.

42. If your car leaves the track, do not attempt to run her back with the current until you are sure she can roll freely without jamming. Movement of the switch when the wheels cannot turn, or are not turning freely, is likely to cause trouble both with the motor and with the switch.

43. Avoid carelessness. Do not allow any metal, viz., your oil can, etc., to touch brass screws on motor boards unless the trolley is off the wire. Do not handle the screws unless the trolley is off or your person touches nothing but dry wood.

44. When examining the motor while the car is in motion, face the rear of the car, or so place yourself that any jerk as in sudden stopping will not pitch you into the machinery.

45. Whenever the trolley leaves the wire the conductor should signal the motorman to stop, and then, after replacing the trolley, he should signal to go ahead. The motorman should bring controller to "off" position as soon as the trolley

jumps and keep it there until the conductor signals to proceed.

46. If you notice any loose motion about the trolley, or if it leaves the line frequently, or if, when running fast on a straight track, there is any flashing between the trolley and the wire, report the same at once.

47. Remember that the trolley wheels need oiling. This should be done as often as necessary. The oil, especially in cold weather, should be of a quality that will not become gummy or sticky.

48. Watch your track joints when going toward a station. If there is sparking ahead of you at the joints, the rail, connections are broken. If the car suddenly gathers speed after passing any point, or if when the lights are on they become very dim and then suddenly brighten up, a broken or loose track connection is also indicated. Report the fact and place promptly.

49. Observe carefully whether the car takes her natural speed for all positions of the switch, and if not report trouble, or, better still, find it yourself and correct, if possible.

50. If motor or car seems to work hard, feel of bearings; if one or more are hot, apply oil and watch frequently; if heat increases, the car should be run in and inspected.

51. If journals squeak, it means that they are running dry and require oil.

52. When storing your car in the house for the night, remove the trolley from the wire, cut out the safety switch, turn the reverse lever so that the car will be ready to run out. Take off the levers and place them on the hook in the office provided for them and marked by the number of the car.

CHAPTER XXIII.

INSTRUCTIONS TO INSPECTORS AND SUPERINTENDENTS.

Upon you depends, more than upon anyone else, the success or failure of any electrical system that is placed in your hands, provided, of course, the system in question is a fairly good one. Cases are not unknown where a given electric system has proved a failure under one superintendent and made to work satisfactorily in every way by his successor. The reverse is also often the case : that a success has been changed to failure by the transfer of the management from the hands of a competent man to those of one unqualified for this most important position. The first requisite in an inspector or superintendent is that he shall be familiar with every detail of the system with which his cars are equipped. If it is important that the motorman shall be familiar with his apparatus, it is much more important that his superiors, from whom he receives instruction, should know it also and in a broader sense. You must remember that the motorman is largely what you make him. He comes to you, perhaps, an untrained hand ; perhaps, what is still worse, he has had instruction on another line by a superintendent who was incompetent or careless, and has thereby acquired habits which he must first unlearn.

You must realize that as you are responsible to

your company for the proper working of the road, so you must hold those under you responsible to yourself. Your reputation and success are therefore largely dependent upon the fidelity and ability of those whose training is in your care. Appreciating this, you will first train yourself. Procure from the company whose apparatus you are using full instructions as to the handling of their machinery, together with blue prints showing the wiring and connections of each individual circuit connected with the electrical equipment of your system. If not an electrician yourself, you should call in someone who is, and who can and will go over with you every detail of your system, and explain the whys and wherefores of everything connected with your plant. Try to understand everything intelligently, and then try your hand at imparting this knowledge to the motormen, in whose hands your reputation and success so largely lie. It is the experience of those who have tried it that the best way to learn a thing accurately one's self is by trying to teach others what you partly know yourself. This is true in mathematics, it is true in science in general and it is true in the car stables, and the superintendent who tries most to impart knowledge to those under him will in a short while outstrip in knowledge him who perhaps at first knew more, but who through a mistaken idea of the dignity or importance of his position has kept his knowledge to himself.

The various means by which he himself has gained his knowledge should as far as possible be placed at the disposal of his employees, and they should be encouraged in every way to avail themselves of them to the fullest extent. In fact, a motorman who will not avail himself of such opportunities when offered is an unsafe man to trust with a car. On some roads the plan has been successfully adopted of furnishing each motorman a diagram of the connections that are being made as the handle of the controlling lever is moved from point to point. On this diagram are also given a few of the most important instructions, and these instructions are supplemented by verbal ones from time to time as occasion requires. Some roads also provide a reading room for their employees, where a few standard works of reference on electrical subjects, and the street railway and electrical papers, are kept on file. This is an excellent plan, and another one is to hang upon the wall of the car house or other convenient place large diagrams of the motor connections and wiring circuits of the car. If equipments of more than one system of control or make are used on the line, each should be similarly represented, so that the differences will be apparent to those who are to manage them. These diagrams might with profit be drawn to half size, or even larger, and in order that the different circuits may be the more readily distinguishable, they should be designated by different colors and the various parts of the equipment lettered for ready reference. These should be hung where the motormen are most apt to congregate when not on duty on their cars, so that they may discuss them at their leisure.

A spirit of emulation among the motormen as to the best care of the equipment of their cars and fulfillment of schedule time should be fostered, and to this end a motorman should be kept on the same car as much as possible. That car should, as far as possible, be considered his own, and he should be held responsible for its record. Nothing is so subversive of effective service as the frequent change of assignment of cars. No motorman can be held responsible for the monthly record of a car which has been handled by a dozen others beside himself, nor can he take a personal pride in its condition at the end of the month under such conditions.

Mr. P. P. Sullivan of the Lowell and Suburban Street Railway Company, in speaking on this subject at a monthly meeting of the Massachusetts Street Railway Association, said: "In addition to creating an interest among the men, and in fact to help create such interest, we have prizes for the motormen whose cars have had the best records in point of expense, delays, etc., and in this manner we are also enabled to find out from the regular men who the best relief men are. Motormen are given printed forms which enable them to call the attention of the night foreman to certain things which may appear wrong, and such form is countersigned by such foreman and forwarded to the superintendent. All loss of mileage or taking off of cars is reported directly to the manager's desk, who exacts an accounting for the cause from the superintendent. By following the above methods we have been enabled to adopt a standard of car mile expenses, and the different foremen are given to understand that if the expenses are kept below such a figure they may expect a present at the end of the fiscal year."

He further says: "We assume that a man before taking charge of a car is absolutely ignorant, has no interest in the apparatus, and we aim to teach him, we endeavor to excite his interest and curiosity, so that he will look out for his motors, inquire for certain motors, create a rivalry so that a man will boast of his record; and we have such instances.

"In the car house skilled mechanics are in charge who are held responsible for results; subdivision of duties and labors in relation to parts of machinery as far as practicable is practiced, so as to more readily locate responsibility. The object is that when a car leaves the shop newly equipped such equipment shall be thoroughly done, through the best material and workmanship, and after that time a thorough inspection. Motors, trucks, and cars are numbered, and an official record is then begun, and date and description of any repairs made are kept and comparisons formed and causes sought."

The ideas above suggested will readily commend themselves to everyone who is not penny wise and pound foolish, for experience has already amply demonstrated that good system, good material, good workmanship and skillful employees who feel a genuine interest in their work are none too good for the best results.

But with all this there will be a failure if the cars are turned over to the motorman in anything but the best condition. It must be understood that the motorman is usually *not* a mechanic and he is *not* an electrician, and that if he were either or both he would have but scant facilities for the exercise of his talents while operating his car on the road. He will have done his whole duty if he handles intelligently what you give him and reports all troubles as soon as they arise.

In the training of motormen it is the practice on some roads to put them first in the machine and repair shops, so that by this preliminary experience they may get an insight into the

anatomy of the car equipment and the adjustment of parts to each other. This is certainly an excellent plan where practicable, and another one which is always feasible, but by far too seldom practiced, is to hold what the doctors would call "clinics" over disabled or diseased motors whenever such are brought in. For instance, if a motor is sent to the repair shop for sparking on account of any of the causes before mentioned, as many of the employees as it is possible to gather together should be called to witness the commutator spark while in action and be told the specific cause thereof, and then they should be shown in the repair shop that the diagnosis is correct and how the trouble is remedied. If a motorman has once seen the sparking due to an open coil, he will forever after recognize it. And so with the other troubles—an object lesson such as the one suggested will be worth more than all the verbal instruction you can give.

Regular inspection should be made every day if possible. It does not pay to allow cars to run until they absolutely refuse to run any longer. In no case does the old adage, "A stitch in time saves nine," apply with more force than in this. I cannot, therefore, insist too strongly that repairs be made just as soon as it is discovered that they are required. If the motorman discovers trouble, he should fix it *if he can;* if he can't fix it, he should run the car into the stables at the first opportunity and report. *In no case should he start out* with a car that is not in condition, and it is the especial business of the inspector to see that he does not.

One of the leading manufacturers of street car motors, in his instructions to inspectors, says:

"Let us impress upon you the importance of keeping a careful watch of all nuts, bolts, screws and all wire connections to see that everything is screwed up tight."

Daily examinations of connections of motors, trolleys, lightning arresters, fuse blocks, etc., should be made, and especial attention should be directed to the connection between the wire from the interior of the car and the trolley base to see that it is in good condition.

In addition to the daily inspection at least once a month the car should be run over the pit and the equipment given a most thorough overhauling.

For testing out faults the magneto bell is an instrument of great general utility, and the managers of electric roads will save their inspectors much time and trouble by providing one or more for their use.

CHAPTER XXIV.

LOCATING FAULTS.

When starting a new car, or one whose connections have been changed, or in fact any car that has come from the repair shops, *always* try the motors *one at a time* to see that the revolution of the controller handle moves the car in the same direction with each motor.

To find an open circuit in the car wiring try both controller handles; if one works, the trouble is probably in the other. If neither works, the trouble is probably not in the controllers. If in one controller, throw the canopy switch to "off." Now hold one of the wires from a magneto bell on the iron work of the truck or motors, and touch successively with the other wire, while the handle of the magneto bell is being rapidly turned, the different contact fingers on the controller; if they ring, the ground connection through that finger is all right. If the bell fails to ring through any one of the fingers, you have located the circuit on which the trouble exists. Trace out that circuit and correct the trouble. If all points ring on one controller, pursue the same method with the other. If both are all right, look for a break or loose connection in the wire running from the trolley base to the fuse block, or to the canopy switch.

The wearing parts and those requiring the most careful attention are the following:

COMMUTATORS.

The *color* of the commutator is in itself a pretty good guide as to its condition. As long as it retains a good gloss and is of a chocolate color it probably does not require attention. It should, however, be carefully examined to see that it is clean and true.

To clean it remove the brushes, and then, while the car is running with that motor cut out, use fine sandpaper applied on a block of wood which exactly fits the curvature of the commutator. Never use emery paper, as the emery itself is a conductor of electricity, and particles detached from the paper may find lodgment between the commutator segments and cause future short circuits.

In case of high bars on the commutator sometimes they may be hammered down by placing a piece of wood or leather between the hammer and the block. Never hammer a commutator block directly with the hammer, however, as it is likely to flatten out the surface so that it will extend over the insulating mica between it and the next block, causing a short circuit. Some authorities say that a hammer should never be used at all, and that the high block should be filed down. But whatever the remedy, a commutator that presents either high bars or flats must be absolutely true after treatment, and this can only be insured by turning it down on the lathe.

The lathe is the remedy also for a rough or eccentric commutator or for mica projecting beyond the surface, as well as for unevenly worn commutators. The turning down of a commutator, however, is a nice piece of work, and should only be intrusted to experienced hands. The cutting should not go deeper than is absolutely necessary to remove the difficulty, and should not extend to the outer end of the commutator; a narrow ridge should always be left on the edge. Crocker and Wheeler say: "In turning a commutator in a lathe a diamond-pointed tool should be used, this being better than either a round or square end. The tool should have a very sharp and smooth edge, and only an exceedingly fine cut should be taken off each time in order to avoid catching in or tearing the copper, which is very tough. The surface is then finished by applying a 'dead smooth' file while the commutator revolves rapidly in the lathe." After turning down or sandpapering the spaces between the commutator blocks should be carefully examined for copper dust or other conducting material. If found, such should be removed. The commutator should be carefully lubricated from time to time, great care being taken, however, not to use an excess. A little vaseline is perhaps as good a lubricator as anything for either the commutator or controlling switch.

The brushes should be examined frequently,—once a day is not too often,—and the brush holders should be cleaned whenever found to be dirty. Brush holders should never be permitted to become loose, and the brushes should not be allowed to become too short, for in this case in the adjustment of the holders the springs will not give sufficient pressure for good contact. Brushes should fit curve of commutator perfectly, and new ones should be filed or sandpapered, if necessary, to make a good fit. Brush tips should be kept clean, and should not be allowed to become wedged in the holders.

THE DROP METHOD OF TESTING FOR FAULTS.

While it is entirely beyond my province to discuss the methods of electrical testing at this time, there are one or two simple methods that are so generally available that it seems well to introduce them here.

If a break or bad contact in a circuit is suspected, it may be readily located by what is known as the "drop method." This is founded on the principle that if everything is normal in a circuit the resistances between any two points equidistant on that circuit should be about the same. If one terminal of a galvanometer be fastened to one point of a circuit, and the other be successively applied to other points on the same circuit at equally increasing distances, the needle will register in a circuit which is intact a regular increase of deflection with every successive point touched. The reason for this is that the galvanometer, being connected up in parallel with the conductor, registers the relative resistance of the portion of the circuit tested and that of its own circuit. The latter being fixed, the indications will be greater as the resistance of the circuit measured is greater. If when the galvanometer wire touches a new point on the circuit the increase of deflection is much greater than it should be, it indicates that there is trouble between this point and the last one touched. A common way of applying this method is to fix the two terminals in a handle of some kind, so that their distance is not varied, and then move these two terminals along the circuit. Since the length of wire measured in this case is always the same, the reading of the galvanometer will always remain the same if everything is right. If the deflection increases at any point, the trouble is at once located between the two terminals of the galvanometer. In testing for breaks in the armature coils the two points are applied to adjacent commutator blocks all around the commutator. The deflection of the galvanometer should not vary between any two successive segments, but if it does there is a loose connection or a break somewhere in the coil or connections between the two.

INSULATION TEST.

Another simple test that is not only of great use, but also available to even the least technical if he have but a magneto bell, is the test for insulation resistance. It is very desirable to know that those portions of your apparatus that should be insulated from each other are so insulated, as, for instance, the armature windings from the armature core, or the brush holder from the brushes. The ordinary magneto bell is rated to ring through from 10,000 to 30,000 ohms resistance. If, therefore, one terminal of the bell be connected to each of the parts that should be insulated from each other, as, for instance, the armature shaft and a commutator segment, or the brush and the brush holder, and the bell can then be caused to ring, it indicates either a very poor insulation between the parts or else a bad short circuit. Its failure to ring does not, however, indicate that the insulation is perfect or sufficient, since it merely indicates that the resistance is somewhat greater than 30,000 ohms (if that be the resistance through which it is rated to ring), whereas the insulation resistance between armature coil and core should not be less than 100,000 ohms for every 100 volts used on the circuit. This test should therefore be considered only as a crude one and more as a test for short circuit than as a test for insulation.

A much more reliable test is that known as the *voltmeter test.* This requires a sensitive high resistance voltmeter, such as the Weston. The Weston 150-volt instrument usually has a resistance of its own of about 15,000 ohms. (Its exact resistance is always stated on a certificate pasted inside the case.) Apply the galvanometer first to some circuit or battery having a high electromotive force—say 100 volts—and note the deflection of the needle. Then connect the parts whose insulation resistance is to be tested with this same circuit in series with the voltmeter. If the armature resistance is to be tested, for instance, connect one terminal of the circuit with the armature shaft, and the other with one terminal or binding post of the machine, and note the new deflection. It will be less than before, because an increased resistance (the insulation resistance) has now been placed in *series* with the galvanometer resistance. The insulation resistance will then be found by the equation:

$$\text{Insulation resistance} = \frac{D \times R}{d} - R,$$

in which D is the deflection due to the galvanometer alone: $d =$ the deflection when the machine is in series with the galvanometer, and R is the resistance of the galvanometer or voltmeter. Thus if the circuit employed in testing is 100 volts, then $D=100$. If the second deflection, viz., that through the galvanometer (say 15,000 ohms), plus the insulation of the machine is 1, our equation becomes : Insulation resistance

$$= \frac{100 \times 15,000}{1} - 15,000 = 1,485,000 \text{ ohms}.$$

BEARINGS.

The importance of keeping the bearings in first-class order will, of course, be apparent at once. Be watchful at all times that they do not become too much worn, else the armature is likely to strike against the fields and become ruined. If, on inspection, you find that the clearance between armature and pole pieces is becoming small, put in new bushings at once. Do not wait until the clearance becomes dangerously small. Looseness of the armature bearings can be detected by lifting on the armature first at one end and then at the other. Loose bearings on the main axle may cause gears to break by reason of their being thrown out of alignment. The grease boxes on the motor should be given careful attention, and should be filled every night or morning before the car is sent out. There is no economy in using an inferior lubricant—rather the reverse ; but the best grease is liable to thicken, and before adding fresh the old should be stirred up with a little oil. In removing covers to inspect grease boxes be careful that none of the dirt or sand falls into the boxes. Occasionally all the grease should be removed, the boxes thoroughly washed with gasoline, and a small quantity allowed to run through the bearings to cut any grease which may have found lodgment there and hardened.

GEARS AND PINIONS.

The life of the gears and pinions depends very largely upon the intelligent care they receive. Proper lubrication is one of the most largely contributive agencies toward long life, but they should be carefully examined every night to see that they have not become loose on the shaft. Since they usually have tapering seats, they may be tightened by firmly tapping them with a hammer. With this frequent examination there is no danger that they will become unduly worn without your knowledge. Tighten up all the bolts whenever examining them. Drive up the key or feather to insure tightness of gears and pinions, and if these are too much worn throw them away and substitute new ones. The unusual "knocking" noises sometimes heard when the car is in operation are usually due either to worn keys, worn-out gears or to some hard object which has become lodged between the teeth. This knocking sound is a warning that must be heeded at once, and the above causes are the first to be suspected. Investigate, and if the pinion is worn out throw it away and put on a new one *with a new key*. A new pinion will not, however, go well with an old and worn gear. If the gear be but slightly worn and too good to throw away, keep it to go with some slightly worn but equally good pinion, but *don't* allow a new wheel to mesh into an old one.

On fitting new gears note carefully that the teeth mesh properly before putting them into use. This is best done by revolving the armature by hand while the car is jacked up. The fitting of new gears is a nice operation, and should not be intrusted to any but a responsible person.

CONTROLLERS.

The controllers should be thoroughly overhauled every night; the contact rings and fingers cleaned and polished when found to be rough, and a little vaseline rubbed on; the screws that hold the contact rings should be tightened if loose, and the ratchet wheel and pawl at the top of the cylinder, as well as the upper and lower bearings of the cylinder, should be carefully lubricated, taking care, however, that too much lubricant is not used and that it does not run down upon the cylinder. In equipments not employing the usual platform cylinder the equivalent parts beneath the car should receive their appropriate attention. Whenever a part becomes worn it should be replaced. In fact, the directions in regard to the controlling devices may all be summed up thus: See that they are in perfect order every night or morning before the car leaves the stables.

CHAPTER XXV.

TROLLEY WHEELS.

The life of the trolley wheel depends upon the quality of the metal, the number of miles that it travels and the care that it receives. Remember that its speed is enormous. Take a trolley wheel that is 6 inches in diameter, for instance. When the car is running at 8 miles an hour, a 36-inch car wheel will make 4482 revolutions, and we know the necessity of lubrication for this. At the same speed and in the same time a 4-inch trolley wheel will make 126,720 revolutions, and yet we are liable to overlook the necessity for lubrication here. In fact, the trolley wheel is subject to something far worse than frictional wear, viz., sparking. Every time a spark occurs on a trolley it means the combustion of so much copper. Oil the trolley wheel in the barn, therefore, as often during the day as it is practicable. Instruct your motorman to oil it when he has an opportunity. See that the wheel does not wobble or flash badly. If it does, it requires attention. Sometimes a wheel may be traced in the dark by its continuous sparking. It requires attention then surely, for if it is not attended to at once it will become so bad that it will have to be thrown away.

The trolley wheel is a little thing, you may think, and not very expensive to replace. The sooner you get over excusing yourself for inattention to *little* things the sooner you will be competent to fill your position. *Be very careful of little things, and there will be no big things to take care of.* This seems like a platitude, but if I can only impress the truth of this statement upon every superintendent I will have accomplished a very great good to the cause of street railroad practice.

The matter of proper tension of the springs at the base of the trolley is one that I am satisfied is not usually given the proper attention. The trolley wheel is, I believe, in nine cases out of ten pushed *too hard against the wire*. It is better to err on the other side, for if the tension be too slight it will make itself manifest, but if it be too great it gives no evidence of the fact until the damage is done. Remember that the resistance of the contact between wheel and wire is not materially increased by increasing the tension of the spring within the allowable limits.

It is a mistaken idea that an extremely high tension will cause the trolley to keep the wire better than a moderate one. On the other hand, the pressure must be sufficient to keep the wheel at its enormous speed from "tapping"—that is, from jumping momentarily from the wire—which it is apt to do if it happens to be a little eccentric or otherwise runs a little unevenly.

In the earlier days, when we did not know so much about these things as we do now, I remember seeing an electric road that was operated by a superintendent who believed there was much to be gained by a stiff pressure; in fact, he carried the idea to an extreme even for those days. The result was that his trolley lifted the wire a couple of feet as it passed along, and as the car jolted and rocked it set the trolley wire into such violent vibration that the motion imparted by one car to the wire would often actually throw the wire off the trolley wheel of another car some distance either in the rear or in advance. He had, in fact, commenced running with too tight a spring, and had sought to correct the difficulty by tightening it still more, which, of course, only made matters worse. His trolley wheels rapidly wore out, and, worse than this, his hangers were knocked to pieces in a very short time, and once when I was in the car, our own trolley getting off, the pole struck a span wire, not only snapping the trolley pole off, but bringing down the whole overhead structure. This man, finally learning his mistake, went to the other extreme, perhaps, but he became a firm convert to the loose spring.

INCANDESCENT LAMPS.

Remember that your lamps are in series, and that any defect in one is visited upon all the others. If one lamp breaks, the circuit is broken; or if one be not screwed in sufficiently tight to make connection, none of them will burn. If, therefore, your lamps refuse to burn, examine them individually to discover the fault before condemning your lamp circuit. The situation of the car lamp is an exceedingly trying one to fill. In the first place there is the constant jar of the car, which tends to break the filament, and in the second there is the great variation in E. M. F. to which it is subjected. The line current is apt to vary from 450 volts to 520 or more, which is, we will say, a variation of fifteen per cent.

Now experience has proved that an increase of three per cent. in the voltage above that for which the lamp was intended will divide its life by two. For instance, if a 100-volt lamp has a life of 1000 hours when used on a 100-volt circuit, it will only have a life of 500 hours if the pressure is increased to 103 volts.

The illumination given by a lamp varies even more widely with the pressure than does its life. No definite law has as yet been discovered as to this, but we know the temperature increases as the *square* of the current, and as the current will be proportional to the pressure, we may say that the temperature will increase as the square of the pressure; the illumination will, however, vary more widely than this, its variation being estimated by some as being as the fifth power of the pressure. These statements being true, or approximately true, one sees at once the strain to which a street car lamp is put.

If your lamps are already burning at a very high pressure—that is, at a pressure above that for which they were intended, or, in other words, if they are burning with high efficiency—a comparatively small increase in pressure will break them in a short time. The rule is, therefore, in street cars *not* to use a high efficiency lamp; first, because under the conditions the light it will give will vary too widely; and second, because its life must necessarily be a very short one. A low efficiency lamp should always be chosen, for then, with the wide variations of pressure to which it is subjected, the variation in light will be less apparent and its life be much longer. Then, again, be careful that all the lamps on a car are of as nearly the same resistance as possible—exactly the same, if that can be arranged, for it is not enough that the five lamps shall in the *aggregate* absorb 500 volts;

they must each absorb the same fraction of this—100 volts. If one lamp absorbs 110 volts and another but 90, the *average* is maintained, but the one that absorbs 110 volts will burn out much the quicker. Lamps have often been very unjustly condemned simply because, being of high resistance, they have burned out first when put in the same series with other lamps, either of the same make or of another, that were of low resistance.

CONCLUSION.

I cannot better illustrate the part which a competent superintendent can play in the general efficiency of a road or a system than by recalling an incident that occurred some years ago in a Western city. Electric railroads were then comparatively new. In this city there were two street railroad companies—one a large corporation owning nearly all the lines in the city, and the other a small corporation owning but one line, and that a short one not more than $3\frac{1}{2}$ miles long. The large corporation equipped one of its lines with the double trolley system, and the small corporation equipped its line with the single trolley. Both lines were of about the same length, but that of the small corporation abounded in long and steep grades, one of which reached $13\frac{1}{4}$ per cent. at one point; while the other, although by no means level, was a much easier line to operate.

A spirit of rivalry sprang up between the two lines which was heightened by a lawsuit in which the single trolley road was made the defendant (it was one of the telephone cases), and in which the double trolley, though not a party to the suit, furnished much of the testimony for the plaintiff. The double trolley people were called in to prove that the double trolley was better than the single trolley; and, among other arguments, showed that with the double trolley the car was independent of the condition of the track, and could run under circumstances (such as heavy snow) where the single trolley would be unable to get current through its motors. That suit was decided in favor of the parties for whom the double trolley people testified.

Winter came on, and with it a heavy snowfall. The double trolley road had every advantage as to track and system to meet this emergency, but the single trolley road had the more efficient superintendent. There is not a street railroad man in the country who would not know his name if I mentioned it. He kept the snow off his tracks and weathered the storm without stoppage of cars. The double trolley superintendent, less alive to the situation, tried to run on top of the snow, and his whole system was blocked for a day. The fact of the blockade of the double trolley system and the successful weathering of the storm by the single trolley system was telegraphed all over the country and taken by the masses as evidence of the superiority of the single over the double trolley. For that special emergency, at least, the facts were exactly the reverse. The true significance of the circumstance was that the single trolley road had the better superintendent, but the public did not understand it in that way. So you must do much for which you will receive no credit from the public at large; but if you keep your cars going and keep them from wearing out, and do this at a minimum expense, your employers will know it and give you full credit.

INDEX.

Advice to Readers, 28
Air Gap, 59
Alternating Current Dynamo, 85
Alternating Currents, Rectification of, 86, 146
Ammeter, 27, 48, 50, 51, 54
Ampere, 12, 13, 15, 21
Amperemeter (see Ammeter)
Ampere-Turn, 58, 90
Analogy between Electricity and Magnetism, 60
Analogy between Water and Electrical Distribution, 116
Analogy of Leaky Pipe and Magnetic Leakage, 65
Analogy of Lines of Force and Elastic Strings, 67
Analogy of the Dog, 10
Analogy of the Pipe, 13, 15
Analogy of the Resistance of Rusty Pipes and Poor Electrical Conductors, 14, 16, 29
Analogy of the Rotary Fan, 136
Analogy of the Waterfall, 12
Analogy of Waterwheel, 51
Armature Coil, Direction of Current in, 85
Armature Coils, Shifting of, 96
Armature Coil, Reversal of Current in, 80
Armature Coils, Stripping of, 97
Armature, Eddy Currents in, 93
Armature, Heating of, 94
Armatures, Closed Coil, 100, 151
Armatures, Drum, 102
Armatures, Open Coil, 100
Armatures, Ring, 102
Armature, Slotted, 95
Arrangement in Multiple, 109, 112
Arrangement in Parallel, 109, 112
Arrangement in Series, 109, 112
Automatic Block System, 184
Automatic Cut-out, 183
Axis of Commutation, 88, 151

Battery, Cost of Making, 32
Battery, Directions for Making, 32
Battery, Direction of Current in, 34
Battery, Dry, 33
Battery, Sal-Ammoniac, 32
Battery, Salt Water, 32
Battery, Zinc-Carbon, 30
Bearings, 226
Bell, Magneto, 224
Binding Wires, 95
Block System, Automatic, 184
Braking, Electrical, 184
Brass, Zinc, Iron, Rusty Pipes, Resistance offered by, 14, 16
Building up of Dynamo, 130
Circuit, Derived (see Shunt Circuit)
Circuit, Shunt, 53
Clearance, 95
Clinics, Motor, 218
Closed Coil Armatures, 100, 151
Closed Conduit System, 190
Commutated Field Control, 169
Commutation, Axis of, 88, 151
Commutation, Line of, 151
Commutation of Currents, 86
Commutator, 88, 89, 221
Compass, 35, 37
Compound Dynamo, 130
Compromise Poles (see Resultant Poles)
Conductors, Poor, 15
Conduit System, 187
Conduit System, Closed, 190
Conduit System, Love, 188
Conduit System, Siemens & Halske, 188
Consequent Poles, 104
Controllers, 228
Conversion of Electrical into Mechanical Units, 27
Conversion of Motor into Dynamo, 184
Counter Electromotive Force, 153
Counter Electromotive Force, an

Essential Factor in the Power of a Motor, 155
Counter Electromotive Force and Ohm's Law, 154
Counter Electromotive Force and Resistance, 154, 157, 160
Counter Electromotive Force and Speed regulation, 157
Counter Electromotive Force, formerly a Bugbear, 155
Counter Electromotive Force, the Measure of Work, 155, 162
Crosby and Bell, 161
Cross Connection of Coils, 124
Current, Flow of, 13
Currents and Magnets, Mutual Effects of, 38
Currents, Commutation of, 80
Currents, Deflection of Needle by, 38
Cut-out, Automatic, 183
Deflection of Needle by Currents, 38
Demagnetization by Shock, 47
Derived Circuit, (see Shunt Circuit)
Direction of Current in Armature Coil, 85
Direction of Rotation of Electric Motor, 141
Directions to Motormen, Specific, 201
Dog, Analogy of the, 10
Drop, 134
Drop Method of Locating Faults, 223
Drum Armatures, 102
Dynamo, Alternating Current, 85
Dynamo, Building up of, 130
Dynamo Compound, 130
Dynamo, Continuous Current, 80
Dynamo-Electric Principle, 129
Dynamo Over Compounded, 135
Dynamo Series, 130
Dynamo, Shunt, 130
Dynamo, The Reversibility of, 135
Earth a Magnet, 46
Eddy Currents in Armature, 93
Effect of Change of Length and Size of Conductor, 19
Electrical Braking, 184
Electrical Horse Power, 23, 148
Electrical into Mechanical Units, Conversion of, 27
Electrical Pressure, 21, 90
Electric Current and its Properties, 28
Electric Motor, 139
Electric Motor, Direction of Rotation of, 141

Electro and Permanent Magnets Compared, 44
Electro-Magnet, 43, 44, 54, 55, 56
Electro-Magnet, Construction of, 72
Electro-Magnet, Tractive Power of, 74
Electro-Magnetic Induction, 75
Electromotive Force, 21, 23 90
Electromotive Force and Strength of Field, 134
Energy, 22
Example, Test for Insulation, 225
Examples, Given C., L., and E. to find R., 19
Examples, Given C., L., and R. to find E., 20
Examples, Given C. and R. to find H. P., 25
Examples, Given E. and R. to find C., 18
Examples, Given E. and R. to find H. P., 24
Examples, Given H. P. and E. to find C., 26
Examples in Speed Regulation, 163, 166,
Expenses, Standard of Car Mile, 216
Faults, Locating, 220, 223
Feeder Wire, 184
Flow of Current, 13
Flux, Resistance to, 60
Foot Pound per Second, 22, 23
Force, Electromotive, 21, 23, 90
Foucault Currents, 94
Friction, 13
Friction, Heat Caused by, 13
Fundamental Units of Electricity 15
Galvanic Battery, Principle of, 30
Galvanometer, 40, 78, 223
Galvanometer, Convenient Coil for, 41
Galvanometer, Directions for Making, 40
Gearing, Double Reduction, 149
Gearing, Single Reduction, 150
Gearless Motor, 150
Gears and Pinions, 227
Heat Caused by Friction, 13
Heating of Armature, 94
Horse Power, Electrical, 23, 24, 148
Horse Power, Mechanical, 22, 23, 148
Induced Poles, 152
Inertia, 159
Inspectors, Instructions to, 213
Instructions to Inspectors, 213

INDEX. 237

Instructions to Superintendents, 213
Insulation Test, 224
Intra-Mural Railroad, 170
Iron, Annealed, Magnetism of, 43, 44
Iron, Cast, Magnetism of, 45
John Scott Legacy and Medal, The, 184
Johnston-Lundell System, 169
Kilowatt, 24
Lamination of Core, 93
Lamps, High Efficiency, 232
Lamps, Incandescent, 231
Lamps, Low Efficiency, 232
Law of Flow in Multiple Circuits, 132
Law of Heating Effects of Current, 159
Law of Magnetic Flux, 56
Law of Magnetizing Effect of Solenoid, 57
Leakage, Magnetic, 64, 65
Leonard System, 176, 177
Like and Unlike Poles, Mutual Effect of, 38
Line of Commutation (see Axis of Commutation),
Lines of Force, 59, 60, 61, 62, 140
Locating Faults, 220, 223
Lowell & Suburban St. Ry. Co., 216
Magnetic Circuit Closed, 63, 75
Magnetic Curves, 67
Magnetic Leakage, 64, 65
Magnetic Poles, Definition of, 63
Magnetic Pressure, 57
Magnetic Saturation, 62
Magnetism and Current, 72
Magnetism, Change of Direction of, 44
Magnetism, Conductors of, 59, 64
Magnetism, Electro, 43, 44. 54
Magnetism of Soft Annealed Iron, 43, 44, 54
Magnetism of Tempered Steel, 43, 44, 54
Magnetism, Permanent, 43, 44, 54
Magnetism, Residual, 74
Magnets, Multipolar, 104, 121, 125, 127
Magneto-Electric Machine, 127
Magneto Bell, 224
Magnetization Assisted by Shock, 46
Magnetization by Contact, 45
Magnetization by Proximity, 45
Magnetization by the Earth, 46
Magnetization of Soft Iron Nearly Instantaneous, 44, 75

Magnetization of Tempered Steel, Time Required for, 44
Magneto-Motive Force, 57, 58
Magnet, Position of, when free to Move in Horizontal Plane, 37
Magnet without Poles, 64
Management of Street Railway Motors, 191
Measuring Instruments, 48
Mechanical Horse Power, 22, 23, 148
Motor Clinics, 218
Motor, Electric, 139
Motor, Electric, Consumes Volts, 51, 119
Motor, Electric, does not Consume Current, 51, 119
Motor Fails to Start, 199
Motor, Gearless, 150
Motor, Slow Speed, 150
Motor Stops, 199
Multiple Arc Arrangement, 109, 112
Multiple Circuits, Law of Flow in, 132
Multiplying Effect of Coils, 40
Multipolar Magnets, 104, 121, 125, 127
Mutual Effects of Currents and Magnets, 38
Mutual Effect of Like and Unlike Poles, 38
North-seeking Pole, 46
North-seeking Pole in Reality a South Pole, 46
Ohm, 14, 15, 18, 21
Ohm, George Simon, 16
Ohm's Law, 16, 17, 21
Ohmmeter, 53
Open Coil Armatures, 100
Overwork, What Constitutes, 99
Permanent and Electro Magnets Compared, 44
Permanent Magnet, 43, 44, 54, 55
Perry System, 181
Pinions and Gears, 227
Pipe, Analogy of, 13, 15
Polarity Dependent on Direction of flow of Current in Solenoid, 55
Polarity, Effect on, of Direction of Flow of Current,
Polarity Independent of Direction of Solenoid, 55
Polarity, Reversing by Reversing Connections, 72
Polarity, Rule for Determining, 71
Poor Conductors, 15
Potential, Difference of, 52. 53

INDEX.

Pressure, Electrical, 21, 90
Pressure, Magnetic, 57
Principle of Galvanic Battery, 30
Production of Steady Currents, 88
Rate of Flow, 21 (see Ampere)
Rate of Work, 22
Readers, Advice to, 28
Rectification of Alternating Currents, 86, 146
Remarks of P. P. Sullivan, 216
Residual Magnetism, 74
Resistance, 14, 21, 48, 53, 54, 56
Resistance, Measurement of, 53
Resistance offered by Iron, Brass, Zinc, Rusty Pipes, 14, 16
Resistance to Magnetic Flux, 56
Resultant Poles, 151
Reversal of Current in Armature Coil, 80
Reversibility of the Dynamo, 135
Rheostat, 161
Rheostat Control, 169
Rheostat Regulation, 161
Ring Armatures, 102
Saturation, Magnetic, 62
Separately Excited Machine, 128
Series Arrangement, 109, 112
Series Dynamo, 130
Series Parallel Control, 169
Series System, 181
Shifting of Armature Coils, 96
Shunt Circuit, 53
Shunt Dynamo, 130
Slotted Armature, 95
Solenoid, Directions for Making, 34
Solenoid, Magnetism and Property of, 42
Solenoid, Magnetism Due Solely to Current, 43
Solenoid, Magnetization by, 35, 48, 55, 57
Solenoid of one Turn, 49
Solenoid of two Turns, 49
Solenoid of three Turns. 50
Solenoid of any Number of Turns, 50

Solenoid, Polarity Produced by, 36, 55
Solenoid, Properties of, 42, 47, 57
Sparking at the Commutator, 194
Specific Directions to Motormen, 201
Speed Control, Johnston-Lundell, 173
Speed Control, Perry System, 185
Speed Regulation, Requirements of, 161
Standard of Car Mile Expenses, 216
Steady Currents, Production of, 88
Storage Battery, Disadvantages of, 186
Storage Battery, Inherent, 185
Storage Battery Traction, 185
Strength of Field and Electromotive Force, 134
Stripping of Armature Coils, 97
Sullivan, P. P., Remarks of, 216
Superintendents, Instructions to, 213
Surface Contact System, 174
Technical Terms, 9, 15
Tempered Steel, Magnetism of, 43, 44
Test for Insulation, 224
Torque, 144
Translating, Current Characteristics in, 116
Translating Device, 110
Trolley Wheels, 229
Units of Electricity, Fundamental, 15
Volt, 12, 13, 15, 23
Voltage, 21
Voltmeter, 27, 48, 51, 53, 54
Voltmeter, Function of, 53
Waterfall, Analogy of the, 12
Watt, 23, 24
What Constitutes Overwork, 99
Wheatstone Bridge, 53
Work, 148
Work, Rate of, 22, 24, 148

Elementary Electro-Technical Series.

BY

EDWIN J. HOUSTON, Ph.D. and A. E. KENNELLY, D.Sc.

Alternating Electric Currents,	Electric Incandescent Lighting,
Electric Heating,	
Electromagnetism,	Electric Motors,
Electricity in Electro-Therapeutics,	Electric Street Railways,
	Electric Telephony,
Electric Arc Lighting,	Electric Telegraphy.

Cloth, profusely illustrated.   *Price, $1.00 per volume.*

The above volumes have been prepared to satisfy a demand which exists on the part of the general public for reliable information relating to the various branches of electro-technics. In them will be found concise and authoritative information concerning the several departments of electrical science treated, and the reputation of the authors, and their recognized ability as writers, are a sufficient guarantee as to the accuracy and reliability of the statements. The entire issue, although published in a series of ten volumes, is, nevertheless so prepared that each volume is complete in itself, and can be understood independently of the others. The books are well printed on paper of special quality, profusely illustrated, and handsomely bound in covers of a special design.

THE W. J. JOHNSTON COMPANY, Publishers,
253 BROADWAY, NEW YORK.

RECENT PROGRESS
IN
ELECTRIC RAILWAYS.

Being a Summary of Current Progress in Electric Railway
Construction, Operation, Systems, Machinery,
Appliances, etc.

By CARL HERING,

AUTHOR OF

"*Principles of Dynamo-Electric Machinery,*" *etc., etc., etc.*

Cloth, 389 pages, 104 Illustrations. Price, $1.00.

The details connected with electric street railways have become so numerous and the systems so varied that the reader is at a loss when he wishes specific information in regard to many desirable points, which can scarcely be expected, as a rule, in a general treatise on the subject. Hering's "Recent Progress in Electric Railways" is particularly valuable from its treatment of details, and elaborates a number of features that have heretofore received only brief notice in other works,—such as high-speed interurban roads and underground tunnel conduit systems,—while the section on construction and operation is very full, and gives much recent engineering and financial data. The historical notes and statistics on the development of the industry will be found complete and reliable. The hundred or more pages devoted to the consideration of details and recent improvements contain information of the greatest value that otherwise could only be obtained by a laborious search through periodical literature. Here the latest inventions and developments in street-railway motors, apparatus, and fittings are described and illustrated in great detail, thus supplying the omissions from more general treatises.

Copies of this or any other electrical book published will be sent by mail, POSTAGE PREPAID, to any address in the world, on receipt of price.

The W. J. JOHNSTON COMPANY, Publishers,
253 BROADWAY, NEW YORK.

REFERENCE BOOK OF
Tables and Formulas
FOR
ELECTRIC STREET RAILWAY ENGINEERS.
ARRANGED AND COMPILED
By E. A. MERRILL,
Author of "Electric Lighting Specifications for the Use of Engineers and Architects."

Flexible Morocco. Price, $1.00.

To a busy man the value of a reference book depends largely on the facility with which he can get from it the information he desires. In the larger works the labor involved in seeking out information, which perhaps is scattered through several sections and encumbered with examples and explanations already familiar to the engineer, is often exceedingly annoying, especially when many times repeated. It is the object of this reference book to avoid such annoyances and meet a practical need by collecting and arranging in a concise, logical order those tables and formulas which are in constant use by the electrical street-railway engineer in making estimates, ordering material, on construction work, etc. All superfluous examples and explanations have been excluded, as well as unnecessary extensions of formulæ into tables when such extensions consist only in the simplest mathematical processes. Not only has considerable care been taken in selecting and checking material compiled directly, but several original tables and formulæ have been added, especially in the sections on track and overhead-work, which will save many calculations. Furthermore, many tables and formulæ have been extended and modified to meet the conditions imposed in street-railway work. The practical arrangement of the work, its condensed style and convenient form, will recommend it to every street-railway engineer. Every heading is in bold-faced type, which easily catches the eye as one glances over the page, thus materially aiding quick reference, and as a further aid a complete cross-index is added. The book is bound in flexible covers and is of convenient size to carry in the pocket.

Copies of this or any other electrical book published will be sent by mail, POSTAGE PREPAID, *to any address in the world, on receipt of price.*

The W. J. JOHNSTON COMPANY, Publishers,
253 BROADWAY, NEW YORK.

PUBLICATIONS OF
THE W. J. JOHNSTON COMPANY.

The Electrical World. An Illustrated Weekly Review of Current Progress in Electricity and its Practical Applications. Annual subscription.............. $3.00

The Electric Railway Gazette. An Illustrated Weekly Record of Electric Railway Practice and Development. Annual subscription....................... 3.00

Johnston's Electrical and Street Railway Directory. Containing Lists of Central Electric Light Stations, Isolated Plants, Electric Mining Plants, Street Railway Companies—Electric, Horse and Cable—with detailed information regarding each; also Lists of Electrical and Street Railway Manufacturers and Dealers, Electricians, etc. Published annually.... .... 5.00

The Telegraph in America. By Jas. D. Reid. 894 royal octavo pages, handsomely illustrated. Russia, 7.00

Dictionary of Electrical Words, Terms and Phrases. By Edwin J. Houston, Ph.D. Third edition. Greatly enlarged. 667 double column octavo pages, 582 illustrations...................... 5.00

The Electric Motor and Its Applications. By T. C. Martin and Jos. Wetzler. With an appendix on the Development of the Electric Motor since 1888, by Dr. Louis Bell. 315 pages, 353 illustrations........... 3.00

The Electric Railway in Theory and Practice. The First Systematic Treatise on the Electric Railway. By O. T. Crosby and Dr. Louis Bell. Second edition, revised and enlarged. 416 pages, 183 illustrations..... .... 2.50

Alternating Currents. An Analytical and Graphical Treatment for Students and Engineers. By Frederick Bedell, Ph.D., and Albert C. Crehore, Ph.D. Second edition. 325 pages, 112 illustrations................. 2.50

Publications of the W. J. JOHNSTON COMPANY.

Gerard's Electricity. With chapters by Dr. Louis Duncan, C. P. Steinmetz, A. E. Kennelly and Dr. Cary T. Hutchinson. Translated under the direction of Dr. Louis Duncan..................................... $2.50

The Theory and Calculation of Alternating-Current Phenomena. By Charles Proteus Steinmetz 2.50

Central Station Bookkeeping. With Suggested Forms. By H. A. Foster......................... 2.50

Continuous Current Dynamos and Motors. An Elementary Treatise for Students. By Frank P. Cox, B. S. 271 pages, 83 illustrations................ 2.00

Electricity at the Paris Exposition of 1889. By Carl Hering. 250 pages, 62 illustrations. 2.00

Electric Lighting Specifications for the use of Engineers and Architects. Second edition, entirely rewritten. By E. A. Merrill. 175 pages............... 1.50

The Quadruplex. By Wm. Maver, Jr., and Minor M. Davis. With Chapters on Dynamo-Electric Machines in Relation to the Quadruplex, Telegraph Repeaters, the Wheatstone Automatic Telegraph, etc. 126 pages, 63 illustrations.. 1.50

The Elements of Static Electricity, with Full Descriptions of the Holtz and Topler Machines. By Philip Atkinson, Ph.D. Second edition. 228 pages, 64 illustrations...................................... 1.50

Lightning Flashes. A Volume of Short, Bright and Crisp Electrical Stories and Sketches. 160 pages, copiously illustrated............................. 1.50

A Practical Treatise on Lightning Protection. By H. W. Spang. 180 pages, 28 illustrations, 1.50

Publications of the W. J. JOHNSTON COMPANY.

Electricity and Magnetism. Being a Series of Advanced Primers. By Edwin J. Houston, Ph.D. 306 pages, 116 illustrations.............................. $1.00

Electrical Measurements and Other Advanced Primers of Electricity. By Edwin J. Houston, Ph.D. 429 pages, 169 illustrations........ 1.00

The Electrical Transmission of Intelligence and Other Advanced Primers of Electricity. By Edwin J. Houston, Ph.D. 330 pages, 88 illustrations............................ 1.00

Electricity One Hundred Years Ago and To-day. By Edwin J. Houston, Ph.D. 179 pages, illustrated 1.00

Alternating Electric Currents. By E. J. Houston, Ph.D. and A. E. Kennelly, D.Sc. (Electro-Technical Series)................................. 1.00

Electric Heating. By E. J. Houston, Ph.D. and A. E. Kennelly, D.Sc. (Electro-Technical Series)...... 1.00

Electromagnetism. By E. J. Houston, Ph.D. and A. E. Kennelly, D.Sc. (Electro-Technical Series)...... 1.00

Electro-Therapeutics. By E. J. Houston, Ph.D. and A. E. Kennelly, D.Sc. (Electro-Technical Series).. 1.00

Electric Arc Lighting. By E. J. Houston, Ph.D. and A. E. Kennelly, D.Sc. (Electro-Technical Series).. 1.00

Electric Incandescent Lighting. By E. J. Houston, Ph.D. and A. E. Kennelly, D.Sc. (Electro-Technical Series).............................. .... 1.00

Electric Motors. By E. J. Houston, Ph.D. and A. E. Kennelly, D.Sc. (Electro-Technical Series)......... 1.00

Publications of the W. J. JOHNSTON COMPANY

Electric Street Railways. By E. J. Houston, Ph.D. and A. E. Kennelly, D.Sc. (Electro-Technical Series).. $1.00

Electric Telephony. By E. J. Houston, Ph.D. and A. E. Kennelly, D.Sc. (Electro-Technical Series).. 1.00

Electric Telegraphy. By E. J. Houston, Ph.D. and A. E. Kennelly, D.Sc. (Electro-Technical Series).. 1.00

Alternating Currents of Electricity. Their Generation, Measurement, Distribution and Application. Authorized American Edition. By Gisbert Kapp. 164 pages, 37 illustrations and two plates 1.00

Recent Progress in Electric Railways. Being a Summary of Current Advance in Electric Railway Construction, Operation, Systems, Machinery, Appliances, etc. Compiled by Carl Hering. 386 pages, 110 illustrations............................... 1.00

Original Papers on Dynamo Machinery and Allied Subjects. Authorized American Edition. By John Hopkinson, F.R.S. 249 pages, 90 illustrations.. 1.00

Davis' Standard Tables for Electric Wiremen. With Instructions for Wiremen and Linemen, Rules for Safe Wiring and Useful Formulæ and Data. Fourth edition. Revised by W. D. Weaver........... 1.00

Universal Wiring Computer, for Determining the Sizes of Wires for Incandescent Electric Lamp Leads, and for Distribution in General Without Calculation, with Some Notes on Wiring and a Set of Auxiliary Tables. By Carl Hering. 44 pages... .... 1.00

Publications of the W. J. JOHNSTON COMPANY.

Experiments With Alternating Currents of High Potential and High Frequency.
By Nikola Tesla. 146 pages, 30 illustrations.......... $1.00

Lectures on the Electro-Magnet. Authorized American Edition. By Prof. Silvanus P. Thompson. 287 pages, 75 illustrations.......................... 1.00

Dynamo and Motor Building for Amateurs.
With Working Drawings. By Lieutenant C. D. Parkhurst... 1.00

Reference Book of Tables and Formulæ for Electric Street Railway Engineers.
By E. A. Merrill.................................. 1.00

Practical Information for Telephonists.
By T. D. Lockwood. 192 pages..................... 1.00

Wheeler's Chart of Wire Gauges........... 1.00

A Practical Treatise on Lightning Conductors. By H. W. Spang. 48 pages, 10 illustrations. .75

Proceedings of the National Conference of Electricians. 300 pages, 23 illustrations.......... .75

Wired Love ; A Romance of Dots and Dashes. 256 pages... .75

Tables of Equivalents of Units of Measurement. By Carl Hering........................ .50

Copies of any of the above books or of any other electrical book published, will be sent by mail, POSTAGE PREPAID, *to any address in the world on receipt of price.*

THE W. J. JOHNSTON COMPANY,
253 BROADWAY, NEW YORK.

AN ILLUSTRATED WEEKLY RECORD OF ELECTRIC RAILWAY PRACTICE AND DEVELOPMENT.

Established January 1, 1896.

THE ONLY ELECTRIC RAILWAY PUBLICATION IN THE WORLD.

As the only publication in the world devoted to the electric railway industry, and the only journal adequately treating the numerous technical features involved in its modern development and practice, the ELECTRIC RAILWAY GAZETTE aims worthily to represent the activity and progressiveness of the important interests to which it is devoted.

Presenting all the news every week, and describing current improvements and developments immediately upon being brought forward, its pages offer to those engaged in the electric railway field the timely advantages enjoyed in other active and important branches of modern industry.

Subscription in advance, One Year, $3.00,
In the United States, Canada or Mexico;
Foreign Countries, $5.00.

The W. J. Johnston Company,
253 BROADWAY, NEW YORK.

THE PIONEER ELECTRICAL JOURNAL OF AMERICA.

Read Wherever the English Language is Spoken.

The Electrical World

is the largest, most handsomely illustrated, and most widely circulated electrical journal in the world.

It should be read not only by every ambitious electrician anxious to rise in his profession, but by every intelligent American.

It is noted for its ability, enterprise, independence and honesty. For thoroughness, candor and progressive spirit it stands in the foremost rank of special journalism.

Always abreast of the times, its treatment of everything relating to the practical and scientific development of electrical knowledge is comprehensive and authoritative. Among its many features is a weekly *Digest of Current Technical Electrical Literature*, which gives a complete *résumé* of current original contributions to electrical literature appearing in other journals the world over.

Subscription { including postage in the U. S., Canada, or Mexico, } **$3 a Year.**

May be ordered of any Newsdealer at 10 cents a week.

Cloth Binders for THE ELECTRICAL WORLD postpaid, $1.00.

The W. J. Johnston Company, Publishers,
253 BROADWAY, NEW YORK.

www.ingramcontent.com/pod-product-compliance
Lightning Source LLC
Chambersburg PA
CBHW021358230426

43666CB00006B/570